Linear Algebra

Modules for Interactive Learning

Using MAPLE®

Linear Algebra

Modules for Interactive Learning Using MAPLE®

PRELIMINARY VERSION

by The Linear Algebra Modules Project
(LAMP)

Eugene A. Herman
Grinnell College

Michael D. Pepe
Seattle Central Community College

Robert T. Moore
University of Washington

James R. King
University of Washington

ADDISON-WESLEY

An imprint of Addison Wesley Longman, Inc.

Reading, Massachusetts • Menlo Park, California • New York • Harlow, England
Don Mills, Ontario • Sydney • Mexico City • Madrid • Amsterdam

Reproduced by Addison-Wesley Publishing Company Inc. from camera-ready copy supplied by the authors.

ISBN 0-201-61838-9

2 3 4 5 6 7 8 9 10 VG 010099

PREFACE

Linear Algebra: Modules for Interactive Learning Using Maple® is a collection of twenty-seven Maple worksheets (i.e., "modules") produced by the Linear Algebra Modules Project (LAMP). Taken together, these modules cover the entire introductory linear algebra course as taught at most colleges and universities. The modular structure of the collection is designed to permit a variety of styles of use under many different teaching conditions. In designing such a flexible set of materials, we had the advantage of working at three very different types of institutions with different types of students, computing facilities, and institutional constraints.

Goals

Our primary objective is to promote more active learning by students in our linear algebra classes. In the standard linear algebra class, we have observed, few students find the material genuinely engaging. The computations are not interesting enough to stimulate students intellectually; the theory often strikes them as too abstract or disconnected from their needs; and the applications are often too sketchy to make an impression. Our goal, therefore, has been to engage students in constructing an understanding of the material. We offer them material that is more concrete, familiar, attractive, and useful to them, so they feel encouraged to participate in the learning process.

Other important goals are related to our use of geometry, our approach to applications, and our attention to development of higher-order thinking.

- **Geometry and Intuition** Our modules use geometry extensively to help students develop their intuition about the concepts of linear algebra. We have found, for example, that once students have a clear understanding of the linear geometry of R^2 and R^3, they become much more at ease with discussions of n-space. Also, by interacting with the visualizations we present and by creating pictures of their own, they become much more confident that their geometric intuition is solidly grounded.

- **Applications** Each application module is substantial enough that students can develop a solid understanding of how linear algebra is used in that application and what significance that contribution has. Since the students have the power of Maple at hand, they can carry out realistic computations and can bring the results to a satisfying conclusion. For example, in the study of least-squares solutions, students plot the data and the approximating line or curve and compare the results of different choices of approximating curves.

- **Higher-order Thinking** One reason students in a standard course have difficulty appreciating the power of the abstractions of linear algebra is that they spend much of their time doing low-level computations on the entries of a matrix. We have found that, by providing appropriate tools in Maple, we can ask new kinds of questions that encourage students to formulate answers in terms of the higher-level objects that Maple commands act on, such as matrices, vectors, and linear transformations. This stimulates students to think at a higher level of abstraction and to appreciate better the power of the subject.

Teaching Options

The modules can be used in many ways, ranging from a lab-based course to a combined lecture/lab course to lecture demonstrations. We have found benefits from each of these modes of use.

- **Lab-Based Course** Typically, in a lab-based course, the book of modules is the sole textbook for the course. Almost all class meetings take place in a computer lab, where students work through the modules on line. The instructor circulates among the students answering questions and engaging in brief discussions with individuals or small groups. Although one generally avoids lecturing in this mode, we recommend holding regular discussions designed to help students see the big picture.

- **Lecture/Lab Course** In this combination of traditional lectures with frequent or occasional labs, the modules are the lab materials and instructors choose which modules to use. Some modules, such as the application modules, work best as supplements to the lectures; others, such as modules that introduce new topics, work best as replacements for the traditional lecture.

- **Lecture Demonstrations** An instructor who wants to use a module but is short of lab time can demonstrate a module in a lecture class. This works best if students have already used some modules so they know what to expect. The instructor can encourage active participation by asking students to suggest what to type into the worksheet and by leading a discussion of the material.

Using the Modules

Each module is divided into two main parts, the Tutorial and the Problems.

- **Tutorial** The Tutorial is further divided into sections and consists of interlaced text (usually brief), examples and demonstrations, and exercises (with answers provided in closed sections). Students work through the Tutorial with some care (yes, they really do!) before going on to the Problems. The Tutorial also serves as a reference for later review. In fact, students often keep their own worked-through copy of the Tutorial as their annotated reference.

- **Problems** The Problems are all intended to be fairly substantial, as they provide the work that students will be graded on. Problems include explorations, applications, constructions (e.g., of specified types of matrices or specified types of pictures or animations), counter-examples, short essays, proofs, true/false questions, and, of course, many challenging computations. Some Problems require pencil and paper computations, since we want our students to know how to perform basic computations without the computer.

Each module begins with a brief statement of purpose, a list of prerequisite concepts, and a summary of the Maple commands used in that module. At the end of each module is a small collection of "Pencil and Paper Practice Questions" which provides students with straightforward review questions to check their grasp of basic concepts and their ability to perform basic computations without the aid of Maple.

The modules are most effective if students work through the Tutorial during the scheduled lab time when the instructor is available to answer questions. Only after students have completed the Tutorial should they begin the Problems, which they should be able to work on primarily during their own time. The instructor decides which Problems to assign, but it is not unusual to assign all or almost all of the Problems.

We always encourage students to read the paper copy of a module before and after the lab and to bring the paper copy to lab. By reading at least the first section of the module before the lab, students will have a head start in lab and will make better use of limited lab time. Reading after the lab will encourage students to reflect on the work they did in lab.

Some instructors encourage students to work together both in going through the Tutorial and in solving the Problems. Others prefer that students solve the Problems individually. We see advantages in each approach. To make grading easier, we suggest that students be required to turn in a printed copy of their solutions to the Problems. To save paper, students can do their work in a blank worksheet or one in which they have copied the assigned Problems.

Software and Hardware Requirements

You will need Release 4 or 5 of Maple V to use our modules. You will also need the LAMP library, which is provided with our modules. This library of Maple functions augments the linear algebra and graphics functions already provided by Maple.

Maple V runs on all the common hardware platforms — Windows PCs, Macintosh computers, and Unix workstations. The LAMP modules and library work in all these environments.

Acknowledgements

We gratefully acknowledge the support of the National Science Foundation, which funded the first two years of our project. We thank our students, who put up with our early efforts, provided crucial feedback, and encouraged us to persevere. We thank the participants in our several minicourses for their suggestions and encouragement. Finally, we thank our editorial and production team at Addison Wesley Longman, which has already provided crucial help, even though our association with them has just begun.

We also acknowledge the many influences that have helped shape our thinking and our strategies. Especially influential have been our involvement with MAX – the MAtriX Algebra Calculator, the ATLAST project (Augment the Teaching of Linear Algebra through the use of Software Tools) headed by

Steven Leon, and the Interactive Mathematics Text Project headed by Gerald Porter and James White. Additionally, we have learned much from the ground-breaking textbooks by David Lay (*Linear Algebra and its Applications, 2nd ed.*, Addison Wesley Longman, 1997), and Gilbert Strang (*Linear Algebra and its Applications, 3rd ed.*, Harcourt Brace Jovanovich, 1976).

Eugene A. Herman
Grinnell College

Michael D. Pepe
Seattle Central Community College

Robert T. Moore and
James R. King
University of Washington

November 12, 1998

CONTENTS

Chapter 5: Vector Spaces

Chapter 6: Eigenvalues and Eigenvectors

Chapter 7: Orthogonality

Linear Algebra Modules Project
Chapter 1, Module 1

Geometric Perspectives On Linear Equations

Purpose of this module

The purpose of this module is to give you experience interpreting linear equations and their solutions geometrically. You will work with systems of linear equations in two and three unknowns, since their solutions can be pictured as points in the plane and in 3-space. The main question to investigate is: What are the possible solution sets for a system of linear equations in 2 or 3 unknowns?

Prerequisites

Familiarity with linear equations in two and three unknowns and their representations as lines and planes, respectively.

Commands used in this module

```
> restart: with(linalg): with(plots): with(lamp):
>
```

Tutorial

Section 1. Linear Equations in Two Unknowns

We begin by looking at a single linear equation in two unknowns:

$$a\,x + b\,y = c$$

This is the equation of a line in the plane, provided the coefficients a and b are not both zero. In Example 1A, we use Maple to solve two such equations simultaneously and to look at their common solutions.

==================

Example 1A: Solve simultaneously the following system of linear equations:

$$x + 4\,y = 6$$
$$3\,x - y = 5$$

Then plot the two lines, and find the location of the solution on the plot.

Solution: We define the Maple variables eqn1 and eqn2 to be the two equations. Then we use the solve(..) command to solve the two equations simultaneously.

```
> eqn1 := x+4*y=6;
> eqn2 := 3*x-y=5;
> solve({eqn1,eqn2},{x,y});
```

So $(x, y) = (2, 1)$ is the unique solution of the given system of two equations. Next we plot the two lines by first solving each equation for y in terms of x.

```
> y1 := solve(eqn1,y);
  y2 := solve(eqn2,y);
> plot({y1,y2},x=-5..5,y=-5..5);
>
```

Note that the two lines intersect at the point (2, 1), as we expected. You can also use the special command drawlines(..) to plot the lines without first having to solve each equation for y:

```
> drawlines([eqn1,eqn2]);
>
```

==================

Exercise 1.1: (a) Replace the second equation above (the one called eqn2) by a new equation so that the system {eqn1, eqn2} has no solutions. But choose the coefficients of x and y so they are not both zero. Check your answer by applying the solve(..) command to your new system.
Maple Note: When you apply Maple's solve(..) command to a system of equations that has no solutions, Maple indicates the absence of solutions by returning no output.
(b) How are your two lines related to one another geometrically? How are their coefficients related to one another algebraically?

Student Workspace

Answer 1.1

Now let's consider three equations in two unknowns. First, let's go back to the original two equations; then we'll consider what can happen when we include a third equation.

```
> eqn1 := x+4*y=6;
  eqn2 := 3*x-y=5;
>
```

Exercise 1.2: Create a third equation (call it eqn3) so the system {eqn1, eqn2, eqn3} has a unique solution, but no two of the three lines are equal.

Student Workspace

Answer 1.2

Section 2. Linear Equations in Three Unknowns

A single linear equation in three unknowns

$$a\,x + b\,y + c\,z = d$$

is the equation of a plane in 3-space, provided the coefficients a, b, and c are not all zero. In Example 2A, we use Maple to solve two such equations simultaneously and to look at their common solutions:

====================

Example 2A: Solve simultaneously the following system of linear equations:

$$x + 2\,y + 3\,z = 3$$
$$2\,x - y - 4\,z = 1$$

Then plot the two planes, and find the location of the solution on the plot.

Solution:

```
> eqn1  := x+2*y+3*z=3;
  eqn2  := 2*x-y-4*z=1;
> solve({eqn1,eqn2},{x,y});
```

So the system of equations has infinitely many solutions: $(x, y, z) = (z + 1, -2\,z + 1, z)$. In this representation of the solutions we refer to z as the "free" variable, because z can have any value and each value of z determines a solution of the system. For example if $z = 4$, then $x = 5$ and $y = -7$, and so $(5, -7, 4)$ is a solution to each of the original two equations. Geometrically, this solution is a parametric representation of the line of intersection of the two planes.

Now let's plot the two planes and look for this line of intersection. As a first step, we solve each equation for z.

```
> z1  := solve(eqn1,z);
  z2  := solve(eqn2,z);
> p1  := plot3d(z1,x=-5..5,y=-5..5,color=blue,style=patchnogrid):
  p2  := plot3d(z2,x=-5..5,y=-5..5,color=green,style=patchnogrid):
  display([p1,p2],scaling=constrained);
>
```

Click on the picture, and then rotate it in 3-space. Try to orient the picture so you are looking right along the line of intersection. (If you find this difficult, use the toolbar and type in the values 127 degrees for Θ and 122 degrees for Φ.) Experiment with some of the other plotting options on the toolbar.

We use Maple's spacecurve(..) command to draw the line of intersection. (We changed the name of the parameter from z to the more traditional t.) Adding this to the previous picture of the two planes provides us with a visual check of our solution.

```
> p3  := spacecurve([t+1,-2*t+1,t],t=-4..4,color=red,thickness=2):
  display([p1,p2,p3]);
>
```

====================

We can obtain this picture with less effort by using the drawplanes(..) command, which is similar to the drawlines(..) command:

```
> drawplanes([eqn1,eqn2]);
>
```

The drawplanes(..) command can also display the planes in another way that is sometimes easier to see:

```
[ > drawplanes([eqn1,eqn2],round=on);
[ >
```

Here's how to interpret this picture. Imagine a sphere with center at the origin and radius 10, and imagine intersecting each plane with this sphere. You then see each of the circles of intersection. You also see the line of intersection (in red) of each pair of planes. Although representing a plane by a circle is somewhat unconventional, this picture is often easier to interpret than the conventional picture of a parallelogram representing a plane when you have several intersecting planes.

```
[ >
```

Exercise 2.1: (a) Replace the second equation above (the one called eqn2) by a new equation so that the system {eqn1, eqn2} has no solution. But choose the coefficients of x, y and z so they are not all zero.

(b) How are your two planes related to one another geometically? How are their coefficients related to one another algebraically?

Student Workspace

Answer 2.1

====================

Example 2B: Now let's consider three equations in three unknowns. First, let's go back to the original two equations; then we'll consider what can happen when we include a third equation.

```
[ > eqn1  := x+2*y+3*z=3;
    eqn2  := 2*x-y-4*z=1;
[ > eqn3  := x+y-z=0;
[ > solve({eqn1,eqn2,eqn3},{x,y,z});
[ >
```

So this system has the unique solution $(x, y, z) = (2, -1, 1)$, and therefore the three corresponding planes must have this single point in common. Look for this point in the two plots below:

```
[ > drawplanes([eqn1,eqn2,eqn3]);
[ > drawplanes([eqn1,eqn2,eqn3],round=on);
[ >
```

Of course, a system of three equations in three unknowns doesn't always have a unique solution, as the next exercise demonstrates.

====================

Exercise 2.2: (a) Replace the third equation above (the one called eqn3) by a new equation so that the system {eqn1, eqn2, eqn3} has more than one solution, but no two of the three planes are equal.

(b) How is your third plane related geometrically to the line of intersection of the first two planes?

Student Workspace:

Answer 2.2

```
[ >
```

Problems

Problem 1: Three equations in two unknowns

Find a system of three linear equations in two unknowns that has no solutions, and choose your equations so that no two of the three lines are parallel. Use {eqn1, eqn2} from Section 1 (given again below) as two of your three equations. Draw the three lines either by hand or using Maple.

```
> eqn1 := x+4*y=6;
  eqn2 := 3*x-y=5;
```

Student Workspace

```
>
```

Problem 2: Three equations in three unknowns

(a) Example 2B gave a system of three linear equations in three unknowns that has a unique solution. In Exercise 2.2, you gave an example of such a system that has an entire line of solutions. We could also give an example of such a system that has no solutions by choosing two of the planes to be parallel, as in Exercise 2.1. In this Problem, you are to find a system of three linear equations in three unknowns that has no solutions but no two of the three planes are parallel. Use {eqn1, eqn2} from Section 2 (given again below) as two of your three equations.

(b) How is your third plane related geometrically to the line of intersection of the first two planes?

```
> eqn1 := x+2*y+3*z=3;
  eqn2 := 2*x-y-4*z=1;
```

Student Workspace

```
>
```

Problem 3: Five equations in three unknowns

Find a system of five linear equations in three unknowns that has a line of solutions, and choose your equations so that no two of the five planes are parallel. Hint: Pick the line of intersection first, and pick one that is easy to work with, such as the z-axis.

Student Workspace

```
>
```

Problem 4: Homogeneous equations

(a) Select any system of two homogeneous linear equations in three unknowns, and choose the equations so the two planes are not parallel. (A linear equation is *homogeneous* if the constant on the right side of the equation is zero; that is, the equation has the form $a\,x + b\,y + c\,z = 0$. This is the equation of a plane that goes through the origin.) Find the solution set of this system and describe it geometrically.

(b) No matter how you chose your two planes in (a), you will always get the same kind of solution set you described in (a). Explain why.

Student Workspace

[>

Problem 5: Essay on solution sets

(a) Describe all the possible solution sets for a system of linear equations in three unknowns. Be careful to include even the most extreme possibilities: The coefficients of x, y, and z might be all zero; some of the equations might be identical to one another; we might have just one equation, or two equations, or three or more equations. Give some illustrative examples.
(b) How does your answer change if all the equations in the system of linear equations are homogenous?

To help you understand the kind of answer that is desired here, we give a sample answer for systems of linear equations in two unknowns:

(a) If we have just one equation, $a\,x + b\,y = c$, there are three possibilities: The solution set is a line, or the solution set is empty (for example, if the equation is $0\,x + 0\,y = 1$), or the solution set is the entire plane (if the equation is $0\,x + 0\,y = 0$). If we have two linear equations, all three of the above solution sets are still possible, and we have the additional possibility that the solution set is a single point. If we have more than two linear equations, all four of these solution sets are still possible.
(b) If all the equations are homogeneous, then the system always has the solution (0, 0). So the solution set is never empty; otherwise, our answer in (a) is unchanged.

Student Writing Space

Pencil and Paper Practice Questions

Note: Maple is not required for any of the problems in this section. Do all of these by hand.

1. Find a system of four equations in two unknowns that has a unique solution but no two of the four lines are equal. (See Exercise 1.2.)

2. Find a parametric representation of the line of intersection of the two planes

$$x + 2\,y + 3\,z = 3$$
$$2\,x - y - 4\,z = 1$$

in which y is the free variable. For two different values of y, determine the corresponding solution of the linear system. (See Example 2A.)

3. Suppose we have a system of three linear equations in three unknowns that represents three planes, and suppose the system has a unique solution. The first two planes cannot be parallel or equal. Why? Therefore they intersect in a line. How is this line related geometrically to the third plane?

Linear Algebra Modules Project
Chapter 1, Module 2

Solving Linear Systems using Maple

Purpose of this module

The purpose of this module is to help you become familiar with some of Maple's many methods for solving a linear system and for reducing a matrix to echelon form. The module will also provide you with practice in solving linear systems without the distraction of doing tedious arithmetic, so you can see the overall strategy more clearly.

Prerequisites

The augmented matrix of a linear system; elementary row operations; row equivalent matrices; Gaussian elimination; pivots; consistent versus inconsistent systems; row echelon form and reduced row echelon form. Some of these prerequisites are reviewed briefly in Section 0.

```
[ >
```

Commands used in this module

```
[ > restart: with(linalg): with(lamp):
[ >
```

Tutorial

Section 0: Quick Review of Prerequisite Concepts and Facts

Definition: Two matrices A and B are ***row equivalent*** if B can be obtained from A by a sequence of elementary row operations. (The elementary row operations are reviewed in Section 2.)

Theorem 1: Suppose we are given two systems of linear equations, both with m equations and n unknowns. If their augmented matrices are row equivalent, then the two systems have exactly the same solution set.

Theorem 2: Every matrix is row equivalent to a matrix in row echelon form.

```
[ >
```

These two theorems underlie the standard method for solving all systems of linear equations. The method consists of the following steps: Write down the augmented matrix A of the linear system. Then

- 1. By using a sequence of elementary row operations, change the augmented matrix A into a matrix R in row echelon form, following the strategy called *Gaussian elimination*. (Gaussian elimination gives us a completely mechanical, and hence readily computerized, method for reducing a matrix to row echelon form, which provides a proof of Theorem 2.)

- 2. Solve the linear system whose augmented matrix is R, following the strategy called *back substitution*. That is, first solve the last equation for its leftmost unknown, then the next to last for its leftmost unknown, etc. (According to Theorem 1, the solutions of this linear system are exactly the solutions of the original linear system.)

```
[ >
```

Section 1: Introduction: Overview of Methods

Sections 2 - 5 of this module are each devoted to a different Maple method for solving a system of linear equations. In each method, we first construct the augmented matrix of the linear system and then reduce it to row echelon form or reduced row echelon form.

The method in Section 2 is closest to the way you would reduce a matrix by hand: We perform step-by-step row reduction using the three elementary row operations. This will give you a chance to practice row reduction without worrying about the accuracy of the individual calculations, which we gladly leave for Maple to carry out.

In Section 3 we move up a level in computation: We use the pivot(..) command to direct Maple to carry out row reduction one column at at time.

Sections 4 and 5 will give you a chance to work with Maple's built-in commands for instantly finding row echelon and reduced row echelon forms of an augmented matrix. Here we will use the gausselim(..) and rref(..) commands.

```
[ >
```

Section 2: Elementary Row Operations

The first step in solving a linear system is to reduce the augmented matrix of the linear system to row echelon form by using elementary row operations.

================

Example 2A: Use elementary row operations to reduce the augmented matrix of the linear system below to row echelon form.

$$-x_1 - 2\,x_2 + x_3 = -1$$
$$2\,x_1 + 4\,x_2 - 7\,x_3 = -8$$
$$4\,x_1 + 7\,x_2 - 3\,x_3 = 3$$

Solution: We begin by entering the augmented matrix for the system.

```
[ > M := matrix([[-1,-2,1,-1],[2,4,-7,-8],[4,7,-3,3]]);
[
```

[>

Maple note: In the matrix(..) command, we enter one row at a time; each row is a list of numbers enclosed in square brackets, and the list of rows is enclosed in another pair of square brackets.

Below are the Maple commands corresponding to the elementary row operations. In each command, M is the name of the matrix we are working on. Take a moment to study the syntax for each of these commands. In particular, take note of the order of the arguments in the addrow(..) command.

addrow(M, i, j, c) : replaces Row j by (c times Row i) + Row j (i.e., $R_j <= c\, R_i + R_j$)

mulrow(M, i , c) : replaces Row i by c times Row i ($c \neq 0$) (i.e., $R_i <= c\, R_i$)

swaprow(M, i, j) : interchanges Row i and Row j (i.e., $R_i <=> R_j$)

In order to keep track of our progress, we will assign a new name, such as *M1*, *M2*, etc., to each matrix created as the result of applying a row operation. Look closely at each command before executing it. Can you predict what it will do?

```
[ > evalm(M);
[ > M1  := addrow(M,1,2,2);
[ > M2  := addrow(M1,1,3,4);
[ > M3  := swaprow(M2,2,3);
[ >
```

Note that matrix *M3* is in row echelon form.

================

The linear system whose augmented matrix is *M3* has the same solutions as our original system.

Exercise 2.1: (By hand) Solve the linear system in Example 2A by converting the matrix *M3* into three equations in the variables x_1, x_2, x_3 . Then use back substitution to find the solution.

Student Workspace

Answer 2.1

[>

Maple can perform back substitution automatically on any augmented matrix in row echelon form. The command to use is backsolve(..). Execute the next line to see how it works on matrix *M3*, and compare the result to your solution.

```
[ > backsolve(M3);
[ >
```

Exercise 2.2 : Using Maple, solve the following system of three equations by applying row operations to its augmented matrix. When you have reduced the augmented matrix to row echelon form, use backsolve(..) to find the solution.

$$x_1 + 5\,x_2 + 2\,x_3 = 1$$
$$-x_1 - 4\,x_2 + x_3 = 6$$
$$x_1 + 3\,x_2 - 3\,x_3 = -9$$

Student Workspace

Answer 2.2

Section 3: The Pivot Command

Maple's pivot(..) command enables you to carry out Gaussian elimination one column at a time. Once you choose the pivot position, Maple applies the appropriate row operations to create zeroes above and below it. Specifically: pivot (M, i, j) treats the (i, j) entry of the matrix *M* as the pivot for the jth column and uses addrow(..) to zero out the remaining entries of the jth column, provided the (i, j) entry is nonzero.

Exercise 3.1: The matrix *M* below is the augmented matrix of the linear system in Exercise 2.2. Use a sequence of pivot(..) commands to reduce *M* to row echelon form. Then use backsolve(..) to solve the system. The matrix and the first pivot command have already been entered in the Student Workspace to get you started.

```
> M := matrix([[1,5,2,1],[-1,-4,1,6],[1,3,-3,-9]]);
>
```

Student Workspace

Answer 3.1

Section 4: Gaussian Elimination: gausselim(M)

Not surprisingly, Maple has a command that carries out Gaussian elimination in one step. The command gausselim(M) reduces *M* to row echelon form. In the next example, we use gausselim(..) to solve a system of linear equations. Two new Maple features will show up in this example. First, you will see how to enter a set of equations into Maple and have Maple find the corresponding augmented matrix. Second, as the solution set in this example turns out to be infinite, we will have to understand how Maple displays this infinite set.

================

Example 4A: Use the gausselim(..) command to reduce to echelon form the augmented matrix of the system of linear equations below.

```
> eqn1 := 2*x[1] - 6*x[2] + 3*x[3] - 2*x[4] = -1;
> eqn2 := -x[1] + 3*x[2] - 2*x[3]  = 4;
> eqn3 := 3*x[1]-9*x[2]+4*x[3]-4*x[4] = 2;
>
```

Note that we use x[1] , x[2] , etc. to get Maple to display subscripted variables x_1, x_2, etc.

Solution: Maple has a command, genmatrix(..), that automatically forms the augmented matrix of a system of linear equations. Execute the next input region to see how it works.

```
> M := genmatrix([eqn1,eqn2,eqn3],[x[1],x[2],x[3],x[4]],flag);
```

Notice that the genmatrix(..) command takes three arguments: a list of the equations, a list of the variables used in those equations, and the word "flag". The word "flag" is required to get an augmented matrix. Execute the next line to see what you get if you leave it out.

```
> C := genmatrix([eqn1,eqn2,eqn3],[x[1],x[2],x[3],x[4]]);
```

Now that we have the augmented matrix, we can use the gausselim(..) command to find its row echelon form.

```
> R := gausselim(M);
>
```

================

Exercise 4.1: (By hand) Answer the following questions about the linear system defined at the beginning of Example 4A by looking at the row echelon matrix *R*:
(a) "The system of equations is consistent and has an infinite number of solutions". Justify this statement by referring to pivot columns.
(b) Which are the leading variables (sometimes called "constrained" variables) and which are free variables?
(c) Express the leading variables in terms of the free variables.

```
>
```

Student Workspace

Answer 4.1

Now let's see what we get from Maple if we apply back substitution to *R*.

```
> soln := backsolve(R);
>
```

Maple note: The solution here is expressed in terms of two parameters, t_1 and t_2. These show up in the right places -- in the second and fourth components, which correspond to the free variables: x_2, x_4. But the names Maple chooses for these variables do not coincide with the original variable names. In particular, t_1 and t_2 correspond to x_4 and x_2, not x_1 and x_2.

If you would like to display the solution in terms of the original variable names, you can use the subs(..) command. Execute the next line and compare its result to the solution you found in Exercise 4.1(c).

```
> subs({t[1]=x[4],t[2]=x[2]},evalm(soln));
>
```

Section 5: Gauss-Jordan Elimination: rref(M)

We now consider the rref(..) command. (The letters "r-r-e-f" stand for "reduced row echelon form.") We begin by applying this command to the augmented matrix that arose in solving the following system in Example 4A :

```
> eqn1 := 2*x[1] - 6*x[2] + 3*x[3] - 2*x[4] = -1;
> eqn2 := -x[1] + 3*x[2] - 2*x[3]          = 4;
> eqn3 := 3*x[1] - 9*x[2] + 4*x[3] - 4*x[4] = 2;
> M := genmatrix([eqn1,eqn2,eqn3], [x[1],x[2],x[3],x[4]],flag);
> R := rref(M);
```

And here, for sake of comparison, is what we got earlier using gausselim(..):

```
> G := gausselim(M);
>
```

Exercise 5.1: (a) (By hand) Compare the echelon matrices R and G above. What do they have in common?

(b) (Maple and/or by hand) Use elementary row operations to reduce G to R.

Student Workspace

Answer 5.1

Section 6: Pivot Columns and Free Variables

====================

Example 6A: Suppose that the augmented matrix for a linear system reduces to the row echelon matrix shown below.

$$\begin{bmatrix} 4 & -3 & 2 & -4 & 10 \\ 0 & 0 & 1 & 2 & -7 \\ 0 & 0 & 0 & 0 & 0 \end{bmatrix}$$

What information does this echelon matrix provide us about the solutions of the corresponding linear system?

Solution: We know that the system is consistent since there is no pivot in the rightmost column. We also conclude that the solution set has two leading variables, x_1 and x_3, corresponding to the pivot columns, and two free variables, x_2 and x_4. The fact that the solution set has two free variables is an instance of the following theorem (see Problem 8):

```
>
```

Theorem 3: Suppose we are given a consistent linear system with m equations and n unknowns. If a row echelon form of the augmented matrix of this system has exactly r nonzero rows (i.e., rows that are not made up entirely of zeros), then the solution set of the system has exactly $n - r$ free variables.

In our example we have $n = 4$ and $r = 2$, so the number of free variables is $4 - 2 = 2$.

====================

Since the leading entry of each nonzero row in an echelon form matrix identifies a pivot column, Theorem 3 can be restated in terms of pivots:

- The number of free variables in the solution set of a consistent linear system equals the number of unknowns minus the number of pivots.

```
>
```

Problems

Problem 1: Solve by row operations and gausselim(..)

Solve the linear system in the next input region twice: First use elementary row operations and then use gausselim(..). Check that both methods give the same result.

```
> eqn1 := 2*x[1] - 2*x[2] - 4*x[3] = -2;
> eqn2 := 2*x[1] - x[2] - x[3] = 2;
> eqn3 := -3*x[1] +5*x[2] +4*x[3] = 3;
> eqn4 := -x[1] - 5*x[2] + 4*x[3] = -3;
>
```

Student Workspace

Problem 2: Solve by pivot(..) and rref(..)

Solve the linear system in the next input region twice: First use pivot(..) and then use rref(..). Check that both methods give the same result.

```
> eqn1 := x[1] + x[2] + 3*x[4] + 3*x[5] = 2;
> eqn2 := -2*x[1] + 2*x[2] - x[3] + 2*x[4] + 3*x[5] = 2;
> eqn3 := 4*x[1] + x[3] + 4*x[4] + 3*x[5] = 2;
> eqn4 := -7*x[1] + 5*x[2] - 3*x[3] + 3*x[4] + 6*x[5] = 4;
>
```

Student Workspace

Problem 3: Consistent or inconsistent?

Use any of our four methods to find out whether the following linear system is consistent or inconsistent. Explain why your conclusion follows from the Maple computation.

```
> eqn1 := 2*x[1] - 6*x[2] + 3*x[3] - 2*x[4] = -1;
> eqn2 := -x[1] + 3*x[2] - 2*x[3]  = 4;
> eqn3 := 3*x[1] - 9*x[2] + 4*x[3] - 4*x[4] = 1;
>
```

Student Workspace

Problem 4: Reasoning from the echelon form of the augmented matrix

The matrix A in the next input region is the augmented matrix of a linear system.

(a) (By hand) Write down the system of linear equations represented by this matrix.

(b) Use gausselim(..) to reduce A, but do not apply backsolve(..) yet. Just use the resulting echelon form to answer the following questions: Is the system consistent? What are the leading and free variables? If the system has a solution, is it unique? Explain your reasoning.

(c) Use back substitution to solve the system.

```
> A := matrix( [ [ 1, -6 , 0 ,0, 5 ] , [ 0, 1, -4, 1, 0] ,
  [ 3 , 2, 1 ,0 , 4 ], [ 3 , 3 , -3 , 1 ,4 ] ]);
>
```

Student Workspace

Problem 5: A system with symbolic constants

The matrix B in the next input region is the augmented matrix of a system of three linear equations in three unknowns. By changing the values of the symbolic constants a and b, we can get not only different solutions but different numbers of solutions. To answer the following question, you will have to apply gausselim(..), not rref(..), to the matrix. For what values of a and b does the system have

(a) no solutions?
(b) a unique solution?
(c) a line of solutions? (i.e., one free variable)
(d) a plane of solutions? (i.e., two free variables)

Be sure to check your answers for a few specific values of a and b by using subs(..) to plug values of a and b into the matrix. The form of subs(..) for a matrix or vector M is:
subs({substitutions}, evalm(M)).

```
> B := matrix([[1,2,a,4],[2,-1,3,b],[3,-4,2,-3]]);
>
```

Student Workspace

Problem 6 (Challenge problem): Why can we solve for some variables but not others?

Maple has a built-in command for solving systems. In the input region below we have entered the system of equations from Example 4A, followed by a command directing Maple to solve the system. Compare the answer to our earlier results.

```
> eqn1 := 2*x[1] - 6*x[2] + 3*x[3] - 2*x[4] = -1;
> eqn2 := -x[1] + 3*x[2] - 2*x[3]  = 4;
> eqn3 := 3*x[1] - 9*x[2] + 4*x[3] - 4*x[4] = 2;
> soln := solve({eqn1,eqn2,eqn3});
```

When using the solve(..) command, we can also direct Maple to solve for particular variables, as in the command below:

```
> soln2 := solve({eqn1,eqn2,eqn3},{x[1],x[4]});
>
```

Notice that now Maple is treating x_2 and x_3 as the free variables. Direct Maple to solve for other pairs of variables. You will discover that there is one pair that it cannot solve for. Explain why. Hint: Try solving for this pair of variables by hand; that may help you identify the difficulty.

Student Workspace

Problem 7: What echelon forms are possible?

(By hand) Does there exist a consistent linear system of three equations and five unknowns with (a) two free variables? (b) three free variables? (c) one free variable? In each of these three cases, either give an example of such an equation, or explain why none can exist by discussing what echelon forms are possible for a 3 by 6 matrix.

Student Workspace

Problem 8: Theorem 3

Explain why Theorem 3 is true. Hint: Ask yourself how many pivot columns and how many non-pivot columns any row echelon form of the augmented matrix must have.

Student Workspace

Problem 9: When $r = m$, the system is consistent

(a) Consider the linear system whose coefficient matrix is the matrix A below and augmented matrix is the matrix M below. (The last column of M is chosen at random.) Use Maple to find the reduced row echelon form of both A and M. Is the linear system whose augmented matrix is M consistent? What are the leading variables and the free variables for the linear system whose augmented matrix is M? Execute the command that defines M several times to see if your answer changes when the last column of M changes.

```
> A := matrix([[5,-1,-6,5,-1],[0,3,3,0,4],[-9,-3,6,-9,-3]]);
> M := augment(A,randmat(3,1));
```

(b) Suppose we are given a linear system of m equations and n unknowns, and suppose that the reduced row echelon form of the coefficient matrix has exactly m nonzero rows. Explain why the linear system must be consistent. (Hint: Looking back at part (a) should help.)

Student Workspace

Pencil and Paper Practice Questions

Note: Maple is not required for any of the problems in this section. Do all of these by hand.

1. For the following linear system

$$-2\,x_1 + 2\,x_2 + a\,x_3 = 1$$
$$-x_1 + x_2 + x_3 = 0$$
$$2\,x_1 - 3\,x_2 - 4\,x_3 = 2$$

(a) Write down the augmented matrix of this linear system, and use Gaussian elimination to reduce it to row echelon form.
(b) For what values of the constant a is the linear system consistent?
(c) For $a = 1$, solve the linear system by using the echelon form you found in (a).

2. The matrix below is an echelon form for the augmented matrix of a linear system.

$$\begin{bmatrix} 2 & 0 & -5 & 1 & 0 & 7 \\ 0 & -1 & 4 & 6 & -3 & 0 \\ 0 & 0 & 0 & 1 & -2 & 1 \end{bmatrix}$$

(a) Simply by inspecting this matrix, determine which are the leading variables and which are the free variables in the solution set.
(b) Solve the linear system by back substitution.

Linear Algebra Modules Project
Chapter 1, Module 3

Some Applications Leading to Linear Systems

Purpose of this module

The purpose of this module is to provide a few examples of the ways in which linear systems arise and are used in practice.

Prerequisites

Familiarity with linear systems and methods of solving them, especially by Gaussian elimination.

Commands used in this module

```
> restart: with(linalg): with(plots): with(lamp):
>
```

Tutorial

Section 1: Fitting a Curve to Data

A common problem in many industries is creating elegant, efficient designs for products. Such designs are usually done on computers these days, using techniques that are known as "computer-aided design" (CAD). Typically, the design is made up of curves and surfaces that the designer manipulates on the computer screen. These curves and surfaces are often made up of graphs of fairly simple functions such as polynomials.

Fitting a Polynomial

Here is a simple design problem:

==================

Example 1A: Find a polynomial of degree 2 whose graph goes through the points

$$(1, 2), (2, 6), (3, 4).$$

```
>
```

Solution: We are looking for a polynomial of the form

$$p(x) = a x^2 + b x + c$$

such that p(1) = 2, p(2) = 6, p(3) = 4. These three conditions will give us three linear equations in the three unknowns *a*, *b*, *c*:

```
> p := x -> a*x^2 + b*x + c;
```

```
> eqn1 := p(1)=2;
  eqn2 := p(2)=6;
  eqn3 := p(3)=4;
```

We solve the system by reducing the augmented matrix of the linear system. Then we substitute the values for *a*, *b*, and *c* back into the polynomial *p*:

```
> A := genmatrix([eqn1,eqn2,eqn3],[a,b,c],flag);
> soln := backsolve(gausselim(A));
> q := subs({a=-3,b=13,c=-8},p(x));
```

Finally, let's graph the polynomial and the original three points together in one plot:

```
> p1 := plot(q,x=0..4):
  p2 := plot({[1,2],[2,6],[3,4]},
   style=point,symbol=circle,color=blue):
  display([p1,p2]);
>
```

==================

Exercise 1.1: Find a polynomial of degree 3 whose graph goes through the points

$$(-1, 20), (1, 12), (2, 5), (3, 16).$$

Graph the polynomial and the four points in a single plot.

Student Workspace

Answer 1.1

Sometimes the slope of the polynomial is also given at selected points; that too leads to linear equations, as you will see in some of the problems.

Fitting a Cubic Spline

If we have a large number of points that we want to fit a curve to, choosing a single polynomial is not wise. For such a large number of points, we would need a polynomial of high degree. But a polynomial of high degree usually has a great many maximums and minimums and therefore may have a very bumpy graph. A better choice of curve may be a *cubic spline*, in which we connect consecutive points with cubic (i.e., degree 3) polynomials. To force adjacent cubic graphs to fit together smoothly, we impose the following "compatibility" condition:

- If the cubics $p(x)$ and $q(x)$ both go through a point (a, b), the first derivatives of p and q must be equal at $x = a$, and their second derivatives must also be equal at $x = a$.

This condition insures that adjoining cubics have the same tangent lines and same concavity at the points where they meet.

```
>
```

Here is a simple example:

==================

Example 1B: Fit a cubic spline to the three points $P = (1, 3)$, $Q = (2, 7)$ and $R = (3, 4)$.

Solution: We need two cubic polynomials, one defined on the interval [1, 2] and one on [2, 3]:

```
> p := x -> a*x^3 + b*x^2 + c*x + d;
  q := x -> e*x^3 + f*x^2 + g*x + h;
```

Notice that there are altogether 8 unknown coefficients in the two polynomials. Our goal is therefore to find a system of 8 linear equations in terms of these unknowns. Solving this system will then give us the coefficients of the two cubics making up the cubic spline.

The polynomial p has to go through the points P and Q. This gives us the first two equations:

```
> eqn1 := p(1)=3;
  eqn2 := p(2)=7;
```

And the second polynomial q has to go through the points Q and R.

```
> eqn3 := q(2)=7;
  eqn4 := q(3)=4;
```

The remaining four equations are based on the first and second derivatives of the polynomials. We can calculate the first derivatives of p and q using Maple's D operator, naming them Dp and Dq , respectively.

```
> Dp := D(p);
  Dq := D(q);
```

In a similar way we define the second derivatives DDp and DDq.

```
> DDp := D(Dp);
  DDq := D(Dq);
```

Since both p and q go through the point $Q = (2, 7)$, they must satisfy the above compatibility condition; that is, the first derivatives of p and q must be equal at $x = 2$, and their second derivatives must be equal there as well:

```
> eqn5 := Dp(2)=Dq(2);
  eqn6 := DDp(2)=DDq(2);
```

So far we have only six equations for our eight unknown coefficients; so we need two more. One convention that is often used to provide the two missing equations is to require that the second derivatives of the spline are zero at the leftmost and rightmost points. So:

```
> eqn7 := DDp(1)=0;
  eqn8 := DDq(3)=0;
```

Now that we have eight equations, we solve the system in the ususal way:

```
> M := genmatrix([eqn1,eqn2,eqn3,eqn4,eqn5,eqn6,eqn7,eqn8],
   [a,b,c,d,e,f,g,h],flag);
> backsolve(gausselim(M));
```

Matching this solution vector with the list of unknowns $[a, b, c, d, e, f, g, h]$ gives us the values for each of the coefficients. We then graph the cubic spline:

```
> cubic1 := subs({a=-7/4,b=21/4,c=1/2,d=-1},p(x));
> cubic2 := subs({e=7/4,f=-63/4,g=85/2,h=-29},q(x));
> p1 := plot(cubic1,x=1..2,color=green):
  p2 := plot(cubic2,x=2..3):
  p3 := plot({[1,3],[2,7],[3,4]},
   style=point,symbol=circle,color=blue):
> display([p1,p2,p3],view=[0..4,0..8]);
>
```

==================

Section 2: Estimating Inside Temperatures From Outside

Suppose we have a thin plate whose temperatures along its outside edges are known, and we want to know the temperatures at its interior points. (We will assume that the temperatures are in a "steady state"; that is, they are not changing with time.)

==================

Example 2A: The rectangular plate below has temperatures of 5 degrees along three edges and 15 degrees along the top edge. Find the temperatures at the four indicated interior points, *t1*, *t2*, *t3*, *t4*.

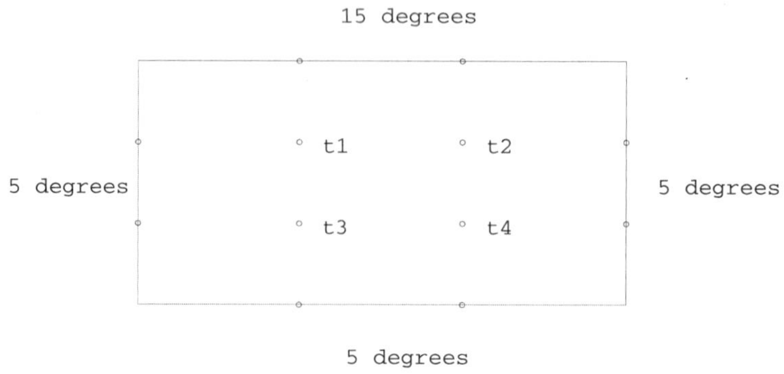

[>

Solution: There are several ways to estimate the temperatures at the interior points, and we will pick the simplest:

- At each interior grid point, assume that the temperature is the average of the temperatures at the four surrounding grid points.

For example, we assume the temperature *t1* is the average of the temperatures 5 (at left), 15 (above), *t2* (at right), and *t3* (below). This method leads to a system of four linear equations in the four unknown interior temperatures:

```
> e1 := t1=(5+15+t2+t3)/4;
  e2 := t2=(t1+15+5+t4)/4;
  e3 := t3=(5+t1+t4+5)/4;
  e4 := t4=(t3+t2+5+5)/4;
> C := genmatrix([e1,e2,e3,e4],[t1,t2,t3,t4],flag);
> soln := backsolve(gausselim(C));
>
```

==================

Of course, when this method is used in a practical problem, one usually wants a very large number of interior points for greater accuracy. So the number of equations and unknowns is then very large.

Problems

Problem 1: Given three points and the slope at each

Find the polynomial of degree five whose graph goes through the points (0, 4), (1, 1), (2, 3) and whose slope at $x = 0$ is -1, at $x = 1$ is 2, and at $x = 2$ is -3. Graph the polynomial and the three points in a single plot.

Student Workspace

Problem 2: Design a ski jump

Design a ski jump that has the following specifications. The ski jump starts at a height of 100 feet and finishes at a height of 10 feet. From start to finish, the ski jump covers a horizontal distance of 120 feet. A skier using the jump will start off horizontally and will fly off the end at a 30 degree angle up from the horizontal. Find a single polynomial whose graph is a side view of the ski jump. [From ATLAST: Computer Exercises for Linear Algebra by Steven Leon, Eugene Herman, and Richard Faulkenberry, Prentice-Hall, 1996, page 8.]

Student Workspace

Problem 3: Cubic spline through four points

In Example 1B we fit a cubic spline to the three points $P = (1, 3)$, $Q = (2, 7)$ and $R = (3, 4)$. Now we extend that problem by adding one more point. Your task is to fit a cubic spline to the three points P, Q, R and the additional point $S = (5, 8)$. So you will have to find three cubics and hence will need 12 equations. The first seven equations that we used In Example 1B are applicable to this problem, but eqn8 is not since 3 is not the rightmost x coordinate. Your task is to come up with equations 8 through 12 and then solve the system.

To get you started, all the relevant commands from Example 1B have been entered in the Student Workspace below as has the third cubic polynomial, $r(x) = j\,x^3 + k\,x^2 + m\,x + n$, which will connect points R and S. Your solution should include a picture that shows the cubic spline along with the four given points.

Student Workspace

Problem 4: Cubic spline on a large data set

(a) Maple has a function that calculates a cubic spline automatically. Apply this function to the points

(0, 0.0), (1, 0.4), (2, 0.4), (3, 0.2), (4, 0.5), (5, -0.5), (6, 0.9), (7, -0.9).

Graph the spline and the points in a single plot. (The commands to do this are given in the Student Workspace. You merely need to execute them.)

(b) Change the sixth point from (5, -0.5) to (5, 0.3), and execute the commands again. Did that alter only the two cubics that go through that point, or did it alter the coefficients of the other cubics as well? Explain.

(c) (By hand) For the problem in (a), Maple had to solve a system of 28 equations in 28 unknowns (seven cubics, each with four unknown coefficients). Write out the second row of the augmented matrix for this system, assuming that the second equation represents the requirement that the graph of the first cubic goes through the second point, (1, 0.4). Suggestion: Denote the 28 unknowns by $a_1, a_2, \ldots, a_{28}$, and the seven cubics by $p_1, p_2, \ldots, p_7$, where

$$p_1 = a_1 x^3 + a_2 x^2 + a_3 x + a_4, \quad p_2 = a_5 x^3 + a_6 x^2 + a_7 x + a_8, \quad \text{etc.}$$

Student Workspace

Problem 5: Temperature grid

Estimate the temperatures at the six interior points of the plate pictured below, using the method of Example 2A.

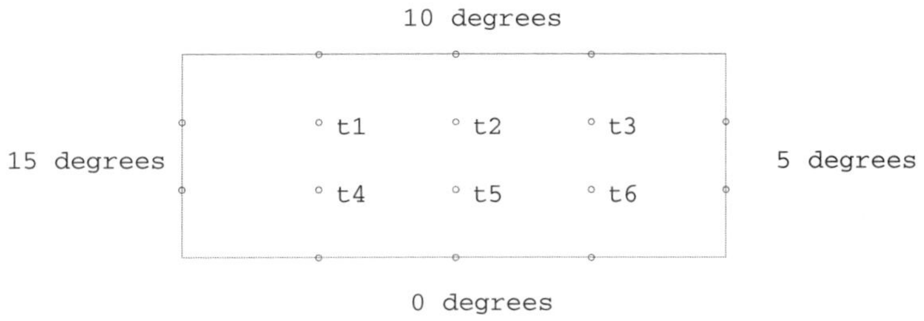

[>

Student Workspace

Pencil and Paper Practice Questions

Note: Maple is not required for any of the problems in this section. Do all of these by hand.

1. (a) When fitting a polynomial to n points in the plane, what degree must the polynomial have? In the resulting linear system, how many unknowns and equations will you have?
(b) When fitting a cubic spline to n points in the plane, how many cubics must you have? In the resulting linear system, how many unknowns and equations will you have?

2. Suppose we are using the method of Example 2A to estimate the interior temperature of a thin rectangular plate from the temperatures along its outside edges, and suppose that our grid of interior points is very large (say 1000 by 1000). Then most of the entries of the resulting augmented matrix of the linear system that represents the four-point averaging method of Example 2A will be zero. What is the largest number of nonzero entries that any row of this matrix can have? What are the only numerical values these nonzero entries can have (ignoring the rightmost column, which depends on the edge temperatures)?

Linear Algebra Modules Project
Chapter 2, Module 1

Geometric Representation of Vectors in R^2

Purpose of this module

The purpose of this module is to help you visualize addition and scalar multiplication of vectors in R^2. We then use these ideas to develop the parametric representation of any line in the plane.

Prerequisites

Some familiarity with vectors.

Commands used in the module

```
[ > restart: with(linalg): with(plots): with(lamp):
[ >
```

Tutorial

Section 1. The Geometry of Vector Addition

Geometric Representations of Vectors

Sometimes we represent a vector $[a, b]$ in R^2 by the point (a, b) in the xy coordinate plane. Other times we represent a vector $[a, b]$ by the "arrow" (i.e., directed line segment) whose tail is at the origin and head is at the point (a, b). In the figure below, you see both the point (–4, 6) (shown as a small blue circle) and the arrow from the origin to the point (-4, 6).

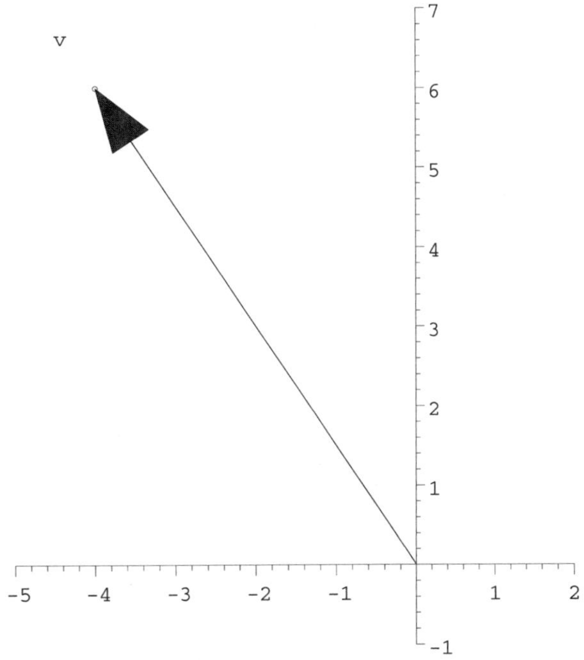

The command drawvec2d(..) draws vectors as arrows. Try it:

```
> drawvec2d([-4,6]);
```

You can also draw any number of vectors at once:

```
> u := colvector([5,-1]);
  v := colvector([-4,6]);
> drawvec2d(u,v);
>
```

Try drawing some other vectors using drawvec2d(..). Also look up help on drawvec2d(..) to see some of its options.

As we proceed through this chapter and beyond, there will be times when it helps to think of vectors as points and other times when arrows provide a more useful visualization. As much as possible, we will try to remind you of which interpretation we are using; but try to get in the habit of considering both possibilities when you approach a new situation.

Vector Addition

Recall that we add vectors *componentwise*; that is, we add the corresponding entries (i.e., components) of the vectors together to get their sum. Below, we define the column vectors u and v, then add them to find $u + v$.

```
> u := colvector([3,1]);
  v := colvector([1,2]);
```

```
evalm(u+v);
>
```

Maple note: Maple does not display the result of a matrix calculation unless you tell it explicitly to evaluate the result using the evalm(..) command as above.

Plotting the two vectors and their sum as arrows provides us with a useful picture of vector addition. Notice that if we form a parallelogram using the two given vectors, then the diagonal of the parallelogram gives their sum:

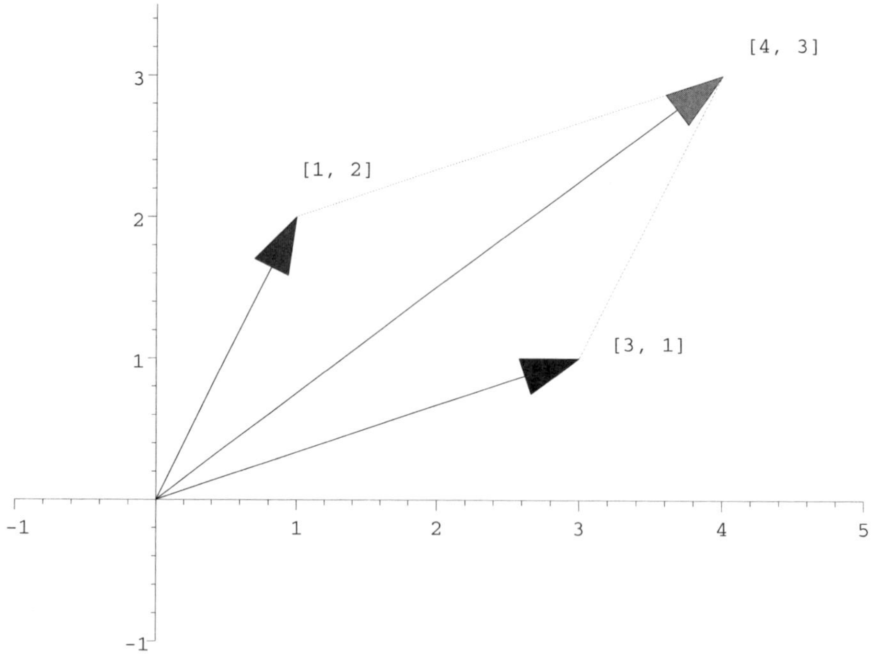

Exercise 1.1: What vector addition is illustrated by the figure below?

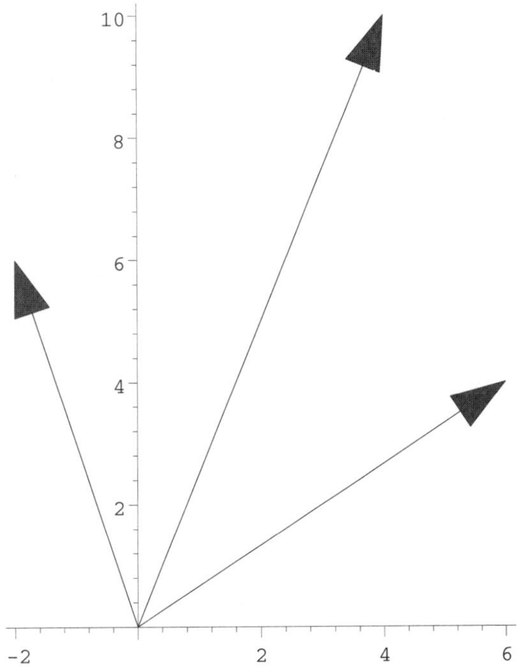

Student workspace

Answer 1.1

Vector Addition as Translation

If you ask a small child to add 8 + 4, most likely they will start at 8 and count: 9,10,11,12. In school their teacher will have them do this on a number line: start at 8 and move 4 places to the right. We can think of vector addition in a similar way. If we start with the vector $v = [1, 2]$ and add to it the vector $u = [3, 1]$, then we move 3 units to the right and 1 unit up and arrive at our result [4, 3]. We say that we have *translated* the vector $v = [1, 2]$ by the vector $u = [3, 1]$, and we refer to u as a *translation* vector.

One way to visualize this is to think of the translation vector $u = [3, 1]$ as if it moves the original point [1, 2] to the resulting point [4, 3] . In the figure below, we move the point [1, 2] to the point [4, 3] along a copy of the vector $u = [3, 1]$ drawn so that its tail begins at the point [1, 2]:

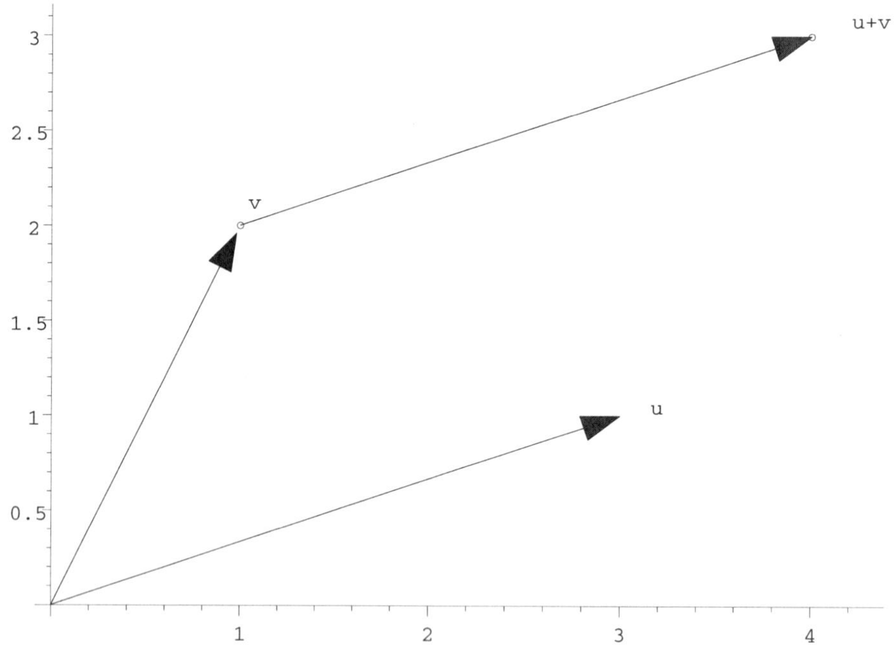

You can also draw this picture using drawvec2d(..):

```
> u := colvector([3,1]);
  v := colvector([1,2]);
> drawvec2d(u,v,[v,u+v]);
>
```

The argument $[v, u+v]$ in the above drawvec2d(..) command describes a vector from the head of v to the head of $u + v$.

Notice that we could just as easily have taken the vector u as our starting point and added the vector v, tail to head, to again arrive at $u + v$. In this case v would be playing the role of the translation vector. Execute the next command to see this other way of getting to $u + v$.

```
> drawvec2d(u,v,[u,u+v]);
>
```

Translation of Geometric Figures

The simple idea of moving a point by adding a vector to it can easily be extended to a set of points. In particular, if the set of points defines a geometric figure, adding a vector to each of the points causes the entire figure to move. For our first example, we will use a collection of four points that form the letter "T". We draw this picture by connecting the four points [2, 1.5], [2, 4], [1, 4], [3, 4] with straight line segments. The columns of the matrix T below contain these four points, and the drawmatrix(..) command draws the points and the line segments connecting them:

```
[ > T := matrix([[2, 2, 3, 1], [1.5, 4, 4, 4]]);
[ > drawmatrix(T);
[ >
```

Let's add the translation vector $p = [3, 1]$ to each of the four defining points, which results in four new points. Hence the entire letter T is moved (i.e., translated) by moving its four defining points the same distance and in the same direction.

```
[ > p := colvector([3,1]);
[ > vectranslate(T,p);
```

For a simpler view of the original and translated points, we use " vectors=off ":

```
[ > vectranslate(T,p,vectors=off);
[ >
```

Exercise 1.2: What translation vector p is required to move the letter T so that its horizontal bar lies on the x axis between 4 and 6? Test your answer by changing the value of p in vectranslate(..).

Student workspace

Answer 1.2

```
[ >
```

Section 2. The Geometry of Scalar Multiplication

Recall that when we multiply a vector by a real number (i.e., a scalar) we multiply *componentwise*; that is, we multiply each entry (i.e., component) of the vector by the scalar. Below, we define a column vector v and a scalar c, then calculate the scalar multiple $c\,v$.

```
[ > v := colvector([3,4]);
    c := 2;
    evalm(c*v);
```

The geometric effect of scalar multiplication is easy to see, if we use arrows to display the vectors v and $c\,v$. In the figure below, the head of the original vector v is in red and the head of the scalar multiple $c\,v$ is in blue.

```
[ > drawvec2d(v,[c*v,headcolor=blue]);
[ >
```

Explore~~~ Try changing the value of c in drawvec2d(..) above. Can you predict what you will get when $c = 3$? $c = -2$? $c = .5$?

```
[ >
```

Exercise 2.1: What do all scalar multiples of a given vector have in common geometrically ?

Student workspace

Answer 2.1

```
[ >
```

Section 3. Parametric Representation of a Line

Parametric Representations of Lines Through the Origin: $t\,v$

If we consider the set of all scalar multiples of a given vector v, we obtain all of the points on the line L through the origin that contains v. It is natural in this context to refer to the vector v as a *direction vector* for the line L. Here is a picture of the line L (in red) along with its direction vector $v = [3, 4]$:

```
> v := colvector([3,4]);
  vectorline(v);
>
```

We can describe all of the points on L algebraically by writing $t\,v$, where t ranges over all the real numbers. The expression $t\,v$ is called a *parametric representation* of the line L, and t is called a parameter. A parametric representation for L is therefore $t\,v = t\,[3, 4] = [3\,t, 4\,t]$. A cartesian equation for L is $y = \frac{4\,x}{3}$.

Exercise 3.1: Find a parametric representation for the line $x + 3\,y = 0$, and give a direction vector for this line.

Student Workspace

Answer 3.1

```
>
```

Parametric Representation of Lines: $t\,v + p$

At the conclusion of Section 1, we observed that if a translation vector p is added to a set of points, then those points move together in the direction of p. So if we are given a line L through the origin, we could take the entire line as our set of points. If we add a vector p to each of the points on the line L, we create a new line $L + p$. If the points on L are described parametrically by $t\,v$, the points on $L + p$ can be described by the expression $t\,v + p$, where t ranges over the real numbers. This expression is called a *parametric representation* of the line $L + p$.

Warning: When we use vectors to represent sets of points, we draw only the heads of the vectors. For example, notice that when drawing the line in Example 3A, we use only the heads of the vectors $t\,v + p$; the shafts of these vectors are not part of the line.

```
>
```

Example 3A: Suppose we are given the line through the origin with direction vector $v = [4, 2]$ and we are given the translation vector $p = [-3, 4]$. Then the translated line has parametric representation $t\,[4, 2] + [-3, 4] = [4\,t - 3, 2\,t + 4]$. The resulting line (shown in blue below) is parallel to the direction vector v and passes through the point at the head of the vector p. Because the head of the vector p is a point on the line, p is sometimes referred to as a *position vector* for the line.

```
> v := colvector([4,2]);
  p := colvector([-3,4]);
  vectorline(v,p);
>
```

Exercise 3.2: Find a cartesian equation for the line $L + p$ above.

Student Workspace

Hint

Answer 3.2

[>

Problems

Problem 1: Vector addition

Six vectors are shown in the figure below. Two of these vectors add up to one of the others. Find this set of three related vectors and write down how they are related, as in B + C = F.

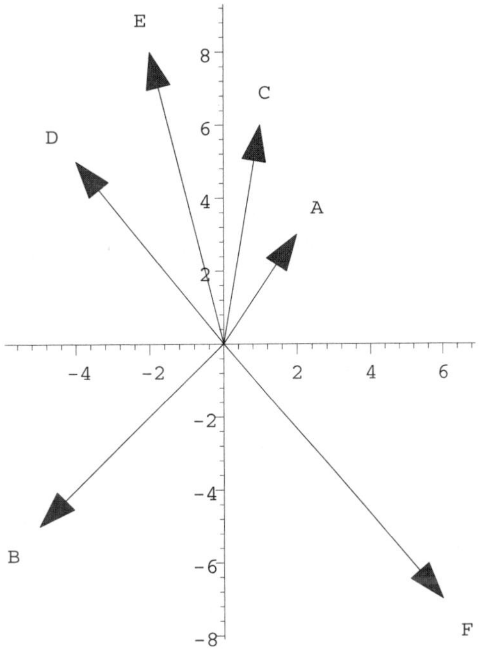

Student Workspace

Problem 2: Position and direction vectors from a picture

Find parametric equations for the red and blue lines shown in the figure below. Find a direction vector *v* for both lines and a position vector *p* for the blue line. Check your work by using vectorline(v,p).

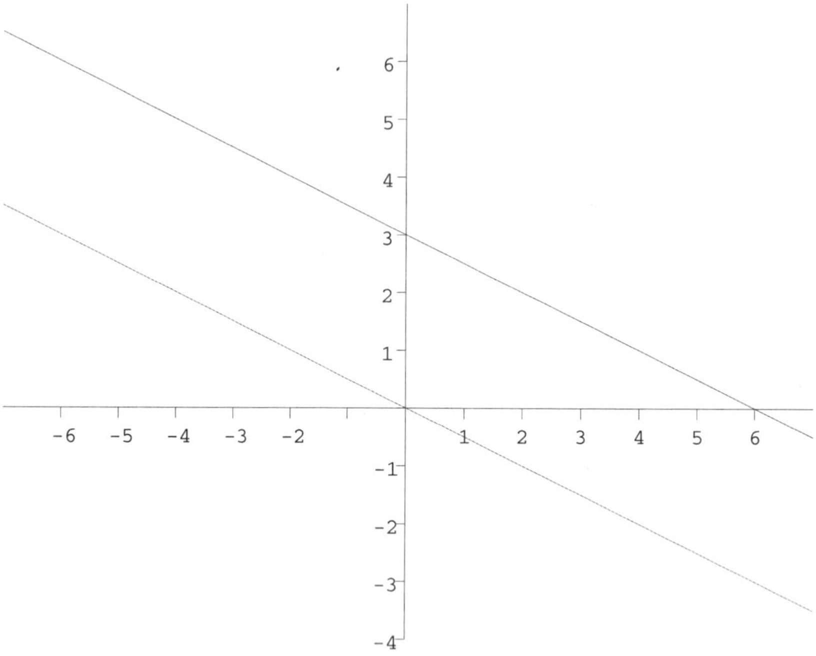

Student Workspace

Problem 3: Parametric equations from a cartesian equation

Consider the line $y = 2x - 5$.
(a) Find a direction vector v for this line.
(b) Find a position vector p for this line.
(c) Find the resulting parametric representation for this line.
(d) There are many answers possible for parts (a) and (b). Choose different answers for both parts and then use them to write another vector representation for the line.
Use vectorline(v,p) to check your work in parts (a) and (b).

Student Workspace

Problem 4: Cartesian equation from parametric equations

Consider the line with parametric form $[3t + 5, -2t + 1]$.
(a) Find a direction vector v for this line.
(b) Find a position vector p for this line.
(c) Find a cartesian equation for this line.
Use vectorline(v,p) to check your work in parts (a) and (b).

Student Workspace

Problem 5: Constructing parallelograms

Use vector algebra (addition, subtraction, and scalar multiplication) throughout this problem.
(a) Find the vertex R of the parallelogram $PQRS$, where $P = (2, 1)$, $Q = (4, 2)$, $S = (1, 5)$ and R is opposite P.
(b) Find the vertices of the translate of the parallelogram PQRS whose center is at the origin. (The "center" of a parallelogram is the common midpoint of its diagonals.)
(c) Find the vertices of the parallelogram that has the same vertex P as in (a), has sides that are twice as long as the parallelogram in (a), and is rotated 180 degrees about the vertex P.
You can check your answers above by using the command polygonplot(..). For example, here is the command to draw the triangle with vertices [2, 1], [4, 2], and [1, 5]:

```
> polygonplot([[2,1],[4,2],[1,5]]);
```

Student Workspace

Pencil and Paper Practice Questions

Note: Maple is not required for any of the problems in this section. Do all of these by hand.

1. (a) Find a parametric representation of the line $y = -2x + 6$. Draw the line and the position and direction vectors on a single plot.
(b) Find another parametric representation of the same line but with a different position vector and a different direction vector.

2. Let $p = \begin{bmatrix} 2 \\ 3 \end{bmatrix}$ and $v = \begin{bmatrix} 4 \\ -1 \end{bmatrix}$. Draw each of the following two sets of points (preferably on graph paper):

$$\{t\,v \mid t \text{ is between -1 and 2}\} \quad \text{and} \quad \{p + t\,v \mid t \text{ is between -1 and 2}\}.$$

3. Use vectors to find the point that is 2/5ths of the way from $P = (1,3)$ to $Q = (6,-7)$. (Check your answer by sketching all three points on graph paper and measuring the distances between them.)

Linear Algebra Modules Project
Chapter 2, Module 2

Linear Combinations of Vectors

Purpose of this module

The purpose of this module is to introduce the concept of linear combinations of vectors. In Section 1, we define linear combinations algebraically, and we introduce the related idea of the span of a set of vectors. In Section 2, we consider the same concepts from a geometric perspective. Finally, in Section 3 we present an example of an applied problem that makes use of linear combinations.

Prerequisites

Vector addition, scalar multiplication, and parametric representation of lines.

Commands used in this module

```
> restart: with(linalg): with(plots): with(lamp):
>
```

Tutorial

Section 1. Linear Combinations of Vectors: An Algebraic Approach

====================

Example 1A: Let $u = \begin{bmatrix} 2 \\ 3 \end{bmatrix}$ and $v = \begin{bmatrix} 4 \\ -1 \end{bmatrix}$. We can create a new vector w by adding scalar multiples of u and v. For example, we can create $w = 3\,u + 2\,v$:

```
> u := colvector([2,3]);
  v := colvector([4,-1]);
  w := evalm(3*u+2*v);
```

Here is the calculation displayed on a single line:

```
> 3*evalm(u)+2*evalm(v)=evalm(3*u+2*v);
```

We say that the vector $\begin{bmatrix} 14 \\ 7 \end{bmatrix}$ has been expressed as a *linear combination* of the vectors $\begin{bmatrix} 2 \\ 3 \end{bmatrix}$ and $\begin{bmatrix} 4 \\ -1 \end{bmatrix}$ with *scalar weights* 3 and 2 respectively. In general, if we want to consider all of the possible linear combinations of the vectors u and v, we write $a\,u + b\,v$ where a and b are any scalar weights (i.e., real numbers).

The set of all linear combinations of the vectors u and v is called the *span* of u and v and is denoted by Span{u,v}. As a result of the calculation above, we now know that the vector $\begin{bmatrix} 14 \\ 7 \end{bmatrix}$ is a member of Span{u,v}.

====================

```
[ >
```

====================

Example 1B: Suppose we were simply given the three vectors $u = \begin{bmatrix} 2 \\ 3 \end{bmatrix}$, $v = \begin{bmatrix} 4 \\ -1 \end{bmatrix}$ and $w = \begin{bmatrix} 14 \\ 7 \end{bmatrix}$. How could we work backwards to find the particular weights for u and v that produce w ?

Solution: We want to express w in the form $a\,u + b\,v$. Thus we have the vector equation

$$a\begin{bmatrix} 2 \\ 3 \end{bmatrix} + b\begin{bmatrix} 4 \\ -1 \end{bmatrix} = \begin{bmatrix} 14 \\ 7 \end{bmatrix}$$

where a and b are the unknowns. Combining terms on the left side of this equation leads us to the familiar problem of solving two linear equations in two unknowns:

```
[ > evalm(a*u+b*v)=evalm(w);
[ > eqn1 := 2*a+4*b=14;
    eqn2 := 3*a-b=7;
```

Reducing the corresponding augmented matrix yields the solution $a = 3$ and $b = 2$:

```
[ > M := genmatrix([eqn1,eqn2],[a,b],flag);
    backsolve(gausselim(M));
[ >
```

Note that the vectors u, v, w are the columns of the matrix M. This suggests a more efficient way to construct the matrix M: Use the command augment(u,v,w), which constructs the matrix whose columns are u, v, w.

```
[ > M := augment(u,v,w);
[ > backsolve(gausselim(M));
[ >
```

====================

Exercise 1.1: Show that the vector k (defined below) is a member of Span{u,v}, where u and v are also defined below.

```
[ > u := colvector([2,3,4]);
    v := colvector([-4,2,1]);
    k := colvector([-40,4,-8]);
```

⊞ Student Workspace

⊞ Answer 1.1

Exercise 1.2: Show that the vector m (defined below) is not a member of Span$\{u,v\}$, where u and v are the vectors defined in Exercise 1.1

```
> m := colvector([20,4,-8]);
```

Student Workspace

Answer 1.2

The examples we have considered so far have all been limited to linear combinations of two vectors. The concepts of linear combinations and span can be applied to any finite set of vectors. Here are the general definitions:

Definition: Given a set of vectors v_1, v_2, ..., v_k in R^n, a ***linear combination*** of these vectors is a vector of the form: $c_1 v_1 + c_2 v_2 + ... + c_k v_k$ (where the scalar weights c_1, c_2, ..., c_k are real numbers). The set of all such linear combinations is called the ***span*** of the vectors v_1, v_2, ..., v_k and is denoted by Span$\{v_1, v_2, ..., v_k\}$.

```
>
```

Section 2. Linear Combinations of Vectors: A Geometric Approach

The purpose of this section is to give you some experience in visualizing linear combinations of vectors. We begin with a picture of our orginal vectors $u = \begin{bmatrix} 2 \\ 3 \end{bmatrix}$ and $v = \begin{bmatrix} 4 \\ -1 \end{bmatrix}$ along with the vector $w = 3\,u + 2\,v = \begin{bmatrix} 14 \\ 7 \end{bmatrix}$ (shown below in green, red, and blue respectively). We have also included a grid based on the vectors u and v. Notice that w is the diagonal of a parallelogram that has vectors $3\,u$ and $2\,v$ as sides.

```
> u := colvector([2,3]):
  v := colvector([4,-1]):
> vectorgrid(u,v,[3,2]);
>
```

Exercise 2.1: The figure below has four points A, B, C and D indicated. Each is a linear combination of u and v; that is, each is of the form $a\,u + b\,v$ for some choice of a and b. Use the grid to help determine the weights for each; then check your work by changing the values for a and b in the command gridgame([a,b]):

```
> gridgame([1,2]);
>
```

Exercise 2.2: In the figure below, a parallelogram is shown in yellow. Express each of the vertices of the parallelogram as a linear combination of the vectors u and v.

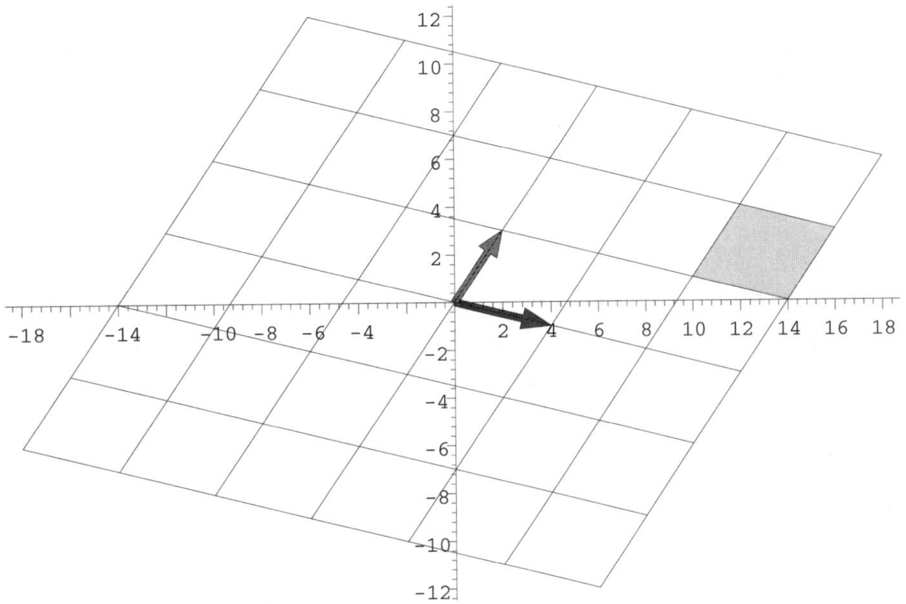

Student Workspace

Answer 2.2

[>

Now that you are able to picture how a particular choice for the scalar weights *a* and *b* produces a particular linear combination *a u* + *b v*, we next will treat *a* and *b* as parameters that can vary. Since we will want to draw many vectors at one time, they will be easier to see if we represent them by points rather than arrows.

=====================

Example 2A: Suppose we let U = { *a u* + *b v* | *a* is in [0, 1], and b is in [0, 1]} . In words: U is "the set of all linear combinations of the vectors *u* and *v* with weights ranging from 0 to 1". Clearly U is an infinite set of vectors, since there is an infinite number of real numbers in the interval [0, 1]. We will refer to U as the ***unit span*** of the vectors *u* and *v*.

Explore~~~

- Imagine that you randomly pick 25 pairs of values for *a* and *b* (all ranging from 0 to 1) and then plot *a u* + *b v* for each. Try to predict what your picture might look like. You can use vectorgrid(..) below to test your prediction. Maple calculates 25 linear combinations *a u* + *b v* by choosing random values for *a* between 0 and 1, *b* between 0 and 1.

```
[ > vectorgrid(u,v,[0..1,0..1],points=25);
[ >
```

- Change the intervals for *a* and *b* in vectorgrid(..). Can you predict what the new picture will be if *a* is between 0 and 2, and *b* is between 1 and 2?

Alternatively, you can use vectorgrid(..) without the option "points=n", which shades in all the linear combinations $a\,u + b\,v$ for *a* and *b* in the designated intervals. For example, here is the unit span of *u* and *v*:

```
[ > vectorgrid(u,v,[0..1,0..1]);
[ >
```

=====================

Exercise 2.3: What range of values for *a* and *b* will produce the set of points shown in yellow in the figure below? Check your answer by using the vectorgrid(..) command.

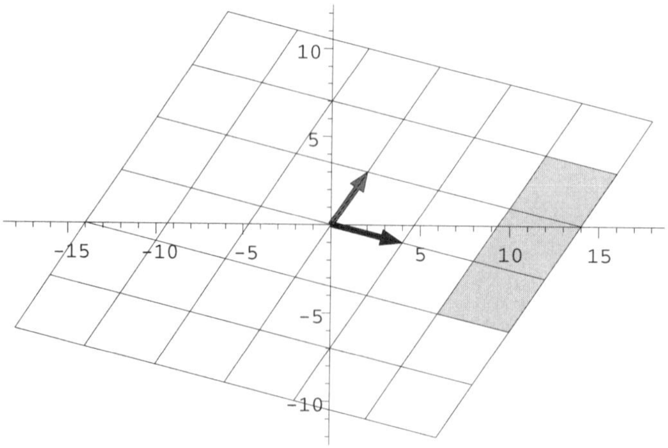


```
[ >
```

Student Workspace

Answer 2.3

Section 3: Application

A coffee shop offers two blends of coffees: House and Deluxe. Each is a blend of Brazil, Kenya, and Sumatra roasts. The percentages for each blend are shown in the table below.

$$\begin{bmatrix} & House & Deluxe \\ Brazil & 30\% & 50\% \\ Kenya & 40\% & 30\% \\ Sumatra & 30\% & 20\% \end{bmatrix}$$

Suppose we make a pounds of the House blend and b pounds of the Deluxe blend. We define two vectors, u and v, based on the columns of the table:

```
> u := colvector([3/10,4/10,3/10]);
  v := colvector([5/10,3/10,2/10]);
```

Exercise 3.1: Suppose the store has 42 lbs of Brazil roast, 34 lbs of Kenya roast, and 24 lbs of Sumatra roast. How much of the House and Deluxe blends should be made in order to completely use up the stock of coffee on hand?

Student Workspace

Answer 3.1

```
>
```

Problems

Problem 1: Linear combinations that span R^2

Show that the vectors u and v below span all of R^2 by showing that for every vector $\begin{bmatrix} x \\ y \end{bmatrix}$ in R^2 the equation $a\,u + b\,v = \begin{bmatrix} x \\ y \end{bmatrix}$ can be solved for a and b.

```
> u := colvector([2,3]);
  v := colvector([4,-1]);
>
```

Student Workspace

Problem 2: Linear combinations that do not span R^2

The following statement is FALSE: "If u and v are any two nonzero vectors in R^2, then every vector in R^2 can be expressed as a linear combination of u and v."

Give an example of nonzero vectors u and v for which the statement is false. Then modify the first part of the statement to make the statement true.

Student Workspace

Problem 3: Seeing linear combinations in R^3

The vectors i, j and k in R^3 are the standard unit vectors [1,0,0] , [0,1,0] , and [0,0,1] respectively.

(a) Give a geometric description for each of the following: Span{ i , j }, Span{ j }, Span{ j , k }.

(b) Is the vector v = [0 , 4 , 5] contained in Span { $i+j$, k }? Give both a geometric and algebraic justification for your answer.

Student Workspace

Problem 4: Redundant linear combinations in R^3

(a) Show that the vector s below is in Span{ u, v, w } and in Span{ u, v, x }, but that s can be expressed in only one way as a linear combination of u, v, w, and in many ways as a linear combination of u, v, x. Find three different sets of coefficients of u, v, x so that their linear combination is s.

(b) Give a geometric explanation for these two different behaviors in (a). Hint: The vector s is actually a linear combination of just u and v, which span a plane in R^3.

```
> u := colvector([1,-1,2]);
  v := colvector([1,2,0]);
  w := colvector([2,2,-3]);
  x := colvector([3,3,2]);
  s := colvector([0,3,-2]);
>
```

Student Workspace

Problem 5: Concrete application (part 1)

[From ATLAST: Computer Exercises for Linear Algebra by Steven Leon, Eugene Herman, Richard Faulkenberry, Prentice-Hall, 1996, page 81.]

Concrete mix, which is used in jobs as varied as making sidewalks and building bridges, is comprised of five main materials: cement, water, sand, gravel, and fly ash. By varying the percentages of these materials, mixes of concrete can be produced with differing characteristics. For example, the water-to-cement ratio affects the strength of the final mix, the sand-to-gravel ratio affects the "workability" of the mix, and the fly ash-to-cement ratio affects the durability . Since different jobs require concrete with different characteristics, it is important to be able to produce custom mixes.

```
>
```

Assume you are the manager of a building supply company and plan to keep on hand three basic mixes of concrete from which you will formulate custom mixes for your customers. The basic mixes have the following characteristics:

	Super-Strong Type S	*All-Purpose* Type A	*Long-life* Type L
cement	20	18	12
water	10	10	10
sand	20	25	15
gravel	10	5	15
fly ash	0	2	8

Each measuring scoop of any mix weighs 60 grams, and the numbers in the table give the breakdown by grams of the components of the mix. Custom mixes are made by combining the three basic mixes. For example, a custom mix might have 10 scoops of Type S, 14 of Type A, and 7.5 of type L. We can represent any mixture by a vector (c,w,s,g,f) representing the amounts of cement, water, sand, gravel, and fly ash in the final mix. The basic mixes can also be represented by vectors:

```
> S := colvector([20,10,20,10,0]);
  A := colvector([18,10,25,5,2]);
  L := colvector([12,10,15,15,8]);
>
```

(a) Give a practical interpretation to the linear combination: $3\,S + 5\,A + 2\,L$.

(b) What does Span$\{S,A,L\}$ represent?

(c) A customer requests 6 kilograms (6000 grams) of a custom mix with the following proportions of cement, water, sand, gravel, and fly ash: 16:10:21:9:4. Find the amounts of each of the basic mixes (*S*, *A*, and *L*) needed to create this mix.

(d) Is the solution unique? Explain.

Student Workspace

Problem 6: Convex combinations

Some linear combinations of vectors provide us with useful ways for describing lines, line segments, and convex polygonal regions in the plane. First, work through Illustrations 1 and 2 below; then do the four parts of Problem 5 that follow. Both illustrations will use the vectors *u* and *v* defined below:

```
> u := colvector([2,4]);
  v := colvector([5,1]);
>
```

Warning: When we use vectors to represent sets of points, we draw only the heads of the vectors. For example, in Illustration 1 we use only the heads of the vectors $a\,u + b\,v$; the shafts of these vectors are not part of the figure.

Illustration 1: Give a geometric description for each of the sets A, B, C, and D defined below. It may be helpful to refer to the figure below (*u* in green , $u + v$ in blue and *v* in red).

(a) $A = \{\, a\,u + b\,v \mid a=0 \text{ and } 0 < b < 1 \,\}$

(b) $B = \{\, a\,u + b\,v \mid b=1 \text{ and } 0 < a < 1 \,\}$

(c) $C = \{\, a\,u + b\,v \mid a=1 \text{ and } 0 < b < 1 \,\}$

(d) $D = \{\, a\,u + b\,v \mid b=0 \text{ and } 0 < a < 1 \,\}$

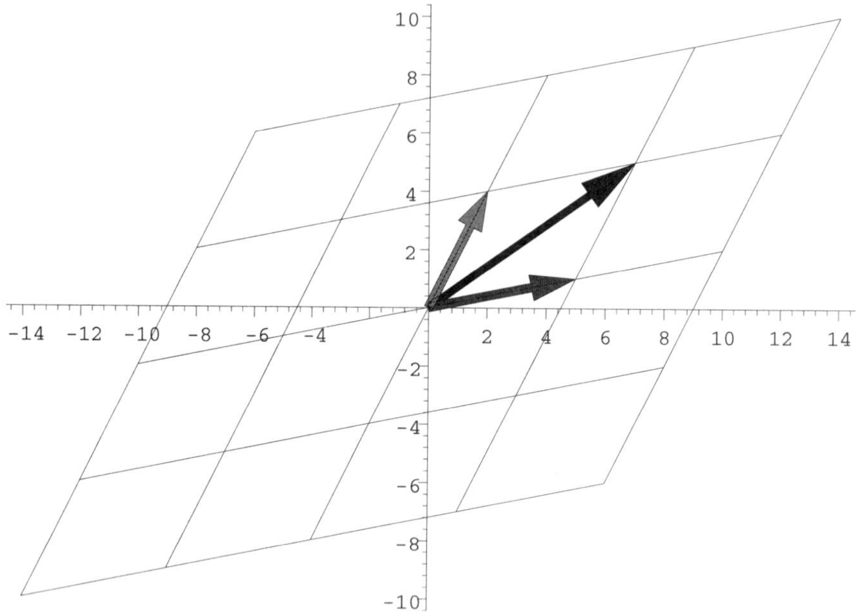

[>

Answer to Illustration 1

[>

Illustration 2: Let M = { $a\,u + b\,v$ | $a = t$ and $b = 1 - t$, where t is any real number}

(a) Give a geometric description of the points contained in the set M.

(b) Give an algebraic justification for your answer to part (a). Hint:

$$t\,u + (1 - t)\,v = v + t\,(u - v)$$

Answer to Illustration 2

[>

Problem 5(a): Let N = { $a\,u + b\,v$ | $a = t$ and $b = 1 - t$, where t is any number in the unit interval [0,1]}.

(i) Describe the three points of N that correspond to $t = 0$, $t = .5$, and $t = 1$. How are they related to the vectors u and v?

(ii) Give a geometric description of all the points contained in the set N.

(iii) Give an algebraic justification for your answer to part (ii).

Student Workspace

[>

Problem 5(b): Consider the two points $A = (2,4)$ and $B = (5,1)$. Describe the set P of all points on and inside the parallelogram $OACB$ (where $C = A + B$ and O = (0, 0)) by using suitable linear combinations of the vectors $u = [2, 4]$ and $v = [5, 1]$.

Student Workspace

[>

Problem 5(c): Consider the two points $A = (2,4)$ and $B = (5,1)$. Describe the set T of all points on and inside the triangle AOB ($O = (0, 0)$) by using suitable linear combinations of the vectors $u = [2, 4]$ and $v = [5, 1]$. Hint: Modify your answer to Problem 5(b).

Student Workspace

[>

Problem 5(d): (Challenge) Consider the three points $A = (2,4)$, $B = (5,1)$ and $C = (8,8)$. Describe the set of all points on and inside the triangle ABC. Your description should express the triangle as a set of linear combinations of the vectors u, v, and w, where $w = [8, 8]$. Hint: Use translation to change this problem into one like Problem 5(c).

Student Workspace

Pencil and Paper Practice Questions

Note: Maple is not required for any of the problems in this section. Do all of these by hand.

1. Let $u = [2, -1]$ and $v = [-1, 3]$. Draw the following set of points, and describe it in words:

$$\{a\,u + b\,v \mid a \text{ is between -1 and 2, and } b \text{ is between 0 and 1}\}$$

2. (a) Suppose u and v are nonzero vectors in R^3 that are not parallel to one another. Describe geometrically Span{ u, v }.

(b) Suppose p and q are two more vectors in R^3, where p is in Span{ u, v } and q is not. Describe geometrically Span{ u, v, p } and Span{ u, v, q }.

Linear Algebra Modules Project
Chapter 2, Module 3

Decomposing the Solution of a Linear System

Purpose of this module

The purpose of this module is to enable you to understand how and why one decomposes the solution set of a linear system so it is expressed in terms of a finite number of vectors. To prepare for the discussion of decomposition, we first investigate vector representations of lines and planes in R^3, as these provide visualizations of the decomposition.

Prerequisites

Vector algebra, vector representation of lines, linear combinations of vectors, span of vectors.

Commands used in this module

```
> restart: with(linalg): with(plots): with(lamp):
>
```

Tutorial

Section 1. Vector Representation of Lines in 3-space

Recall that a line in R^2 can be described parametrically in the form $t\,v + p$, where v is a nonzero *direction vector, p* is a *position vector*, and t is a parameter that ranges over all the real numbers. We picture the line (shown in blue below) passing through the point p and and parallel to the vector v:

```
> v := colvector([-2,4]):
  p := colvector([3,4]):
  vectorline(v,p);
>
```

Recall that we can understand this vector representation by constructing the line in two steps. First form the line through the origin in the direction of v (shown in red). This line is simply all scalar multiples of v and so has the vector representation $t\,v$. Next, translate this line by adding the vector p to all the vectors $t\,v$. The point at the origin moves to p, and the remaining points shift to form the line $t\,v + p$ (shown in blue).

Lines in R^3 can be described in exactly the same vector form, $t\,v + p$.

Warning: When we use vectors to represent sets of points, we draw only the heads of the vectors. For example, notice that when drawing the line in Example 1A, we use only the heads of the vectors $t\,v + p$; the shafts of these vectors are not part of the line.

====================

Example 1A: Find a vector representation of the line in R^3 that passes through the point p = [3,1,-4] and has direction vector v = [2,-5,3].

Solution: Calculate the expression $t\,v + p$. Here's how it looks using Maple:

```
> v := colvector([2,-5,3]);
  p := colvector([3,1,-4]);
  t*evalm(v)+evalm(p) = evalm(t*v+p);
```

Here is a picture of the vectors v and p and the line $t\,v + p$. Try looking at it from different viewpoints, with and without axes displayed. Notice that the lines and the direction vector remain parallel in all views.

```
> v := colvector([2,-5,3]):
  p := colvector([3,1,-4]):
  vectorline(v,p);
> 
```

====================

Maple Note: To draw the line by itself, you can use Maple's spacecurve(..) command. Notice the syntax for this command: The first three entries in the square brackets are the parametric equations and the fourth entry specifies the range for the parameter t to be plotted.

```
> spacecurve([2*t+3,-5*t+1,3*t-4],t=-2..2,
   color=blue,axes=boxed);
> 
```

Exercise 1.1: (a) What are the coordinates of the two endpoints of the portion of the line drawn using the spacecurve command above?
(b) Does the point $[9, -14, 5]$ lie on the line?

Student Workspace

Answer 1.1

Exercise 1.2: Find a direction vector and a position vector for the line that has parametric form

$$[2\,t + 2,\ 3\,t + 5,\ -t - 4]$$

Student Workspace

Answer 1.2

```
> 
```

Section 2. Vector Representation of Planes: $t\,v + s\,w$

In Module 2 we defined the "unit span" for two vectors in R^2. Recall that, for two nonparallel vectors, the set of all points in the unit span filled out the parallelogram determined by the vectors. If we take two nonparallel vectors in R^3, the unit span will again be the parallelogram defined by the vectors. But now this parallelogram sits in three-dimensional space. Execute the next line to see the unit span for two vectors in R^3 :

```
> v := colvector([5,-1,2]);
  w := colvector([3,7,1]);
> unitspan(v,w);
>
```

Explore~~~ Try some other pairs of vectors in unitspan(..). Can you predict the picture you will get for these pairs: [0, 1, 0] and [0, 0, 1], or [1, 1, 1] and [1, 0, 0]? Try to visualize the unit span in each case before creating the picture.

Student Workspace

When we studied Span{v, w} in Module 2, we concentrated on vectors in R^2. Now let's consider Span{v, w} when v and w are vectors in R^3. Recall that Span{v, w} is defined as the set of all linear combinations of the vectors v and w: $t\,v + s\,w$. So instead of limiting the coefficients t and s to the unit interval [0,1] as we do for the unit span, we allow t and s to range over all the real numbers. Imagine the yellow parallelogram expanding, creating a infinite plane; that is Span{v, w}. Since the vectors v and w are parallel to this plane and completely determine it, they are referred to as its *direction* vectors. The plane is then written in the vector form $t\,v + s\,w$, where it is understood that t and s range over all the real numbers.

```
>
```

Exercise 2.1: We have been considering only vectors v and w that are not parallel, because that leads to the most interesting case. But the unit span and Span{v, w} are still well defined for vectors that are parallel. Describe geometrically what each looks like. To make this question more specific, assume that v = [1, 3, 4] and w = [2, 6, 8].

Student Workspace

Answer 2.1

Exercise 2.2: State two nonparallel direction vectors for the yz plane in R^3. To make this exercise a little more interesting, use vectors that are not parallel to the y or z axes.

Student Workspace

Answer 2.2

```
>
```

Since we now know that, when v and w are nonparallel vectors, $t\,v + s\,w$ represents a plane through the origin, we can study planes from the opposite direction: Given a cartesian equation of a plane, find its vector representation.

====================

Example 2A: For the plane $x + 2\,y + z = 0$, which passes through the origin, find two nonparallel direction vectors.

```
> eqn1 := x+2*y+z=0;
```

Solution: We could easily solve this equation by hand. (Do so, and compare with Maple's solution below.)

```
> M := genmatrix([eqn1],[x,y,z],flag);
> soln := backsolve(M,t);
>
```

To find direction vectors for the plane, we factor or "decompose" the above solution into a linear combination of two vectors v and w. Keep in mind that t_1 and t_2 are parameters here, playing the role of t and s in the expression $t\,v + s\,w$.

$$\begin{bmatrix} -2\,t_2 - t_1 \\ t_2 \\ t_1 \end{bmatrix} = t_1 \begin{bmatrix} -1 \\ 0 \\ 1 \end{bmatrix} + t_2 \begin{bmatrix} -2 \\ 1 \\ 0 \end{bmatrix}$$

So $v = [-1, 0, 1]$ and $w = [-2, 1, 0]$. Note that the components of v are just the coefficients of t_1 in the solution, and the components of w are the coefficients of t_2. Another way to find v and w from the above solution is to first set $t_1 = 1$ and $t_2 = 0$ (which produces v) and then set $t_1 = 0$ and $t_2 = 1$ (which produces w). Here is how Maple would do that:

```
> v := subs({t[1]=1,t[2]=0},evalm(soln));
> w := subs({t[1]=0,t[2]=1},evalm(soln));
>
```

Conclusion: The solution set of the equation $x + 2\,y + z = 0$ can be expressed in the form $t\,v + s\,w$, where v and w are as defined above.

====================

There are many alternative pairs of direction vectors that we could have used to describe our plane. In fact, any two nonparallel vectors in the plane will span the plane. We used the procedure above because it demonstrates how the direction vectors arise naturally from the equation of the plane when we "decompose" the solution. This is a process that you will use repeatedly in expressing solutions of linear systems in vector form.

```
>
```

Explore ~~~ We can use plot3d(..) or drawplanes(..) to see the plane $x + 2\,y + z = 0$, and can then use drawvec3d(..) and display3d(..) to add the vectors v and w to this plot. Since the plane passes through the origin, we expect to see the vectors sitting right in the plane. Move the plane around to confirm this.

```
> v := colvector([-1,0,1]):
  w := colvector([-2,1,0]):
> plane := plot3d(-x - 2*y, x=-2..0,y=-1..1,
    axes=framed,style=patchnogrid,color=green):
> vects := drawvec3d([v,label=v],[w,label=w]):
> display3d([plane,vects],axes=boxed);
>
```

Exercise 2.3: (a) Find the vector representation for the plane through the origin that has the cartesian equation $2\,x - y + 4\,z = 0$.

(b) Confirm that your vectors v and w are parallel to the plane by showing that they are orthogonal to a normal vector for the plane.

Student Workspace

Answer 2.3

Exercise 2.4: Find a cartesian equation for the plane through the origin that has direction vectors $v = [-1, 2, 5]$ and $w = [0, 4, 3]$.

Student Workspace

Answer 2.4

[>

Section 3. Vector Representation of Planes: $p + t\,v + s\,w$

If we take a vector representation of a plane through the origin, $t\,v + s\,w$, and add a position vector p, we get $p + t\,v + s\,w$. Geometrically, the result is a translation of the original plane by the vector p to a new plane that now passes through the point p rather than the origin. The original plane and the translated plane are parallel and therefore have the same direction vectors, namely v and w.

=====================

Example 3A: For the plane $x + 2\,y + z = 6$, find a vector representation of the form $p + t\,v + s\,w$. From our discussion in the above paragraph, we know that this problem is closely connected to the problem of finding direction vectors for the parallel plane through the origin, $x + 2\,y + z = 0$. The two solution methods described below make clear this connection.

[>

Solution Method 1:

(a) Find the direction vectors: In Example 2A, we found the direction vectors for the parallel plane through the origin, $x + 2\,y + z = 0$, to be $v = [-1, 0, 1]$ and $w = [-2, 1, 0]$. Therefore our given plane, $x + 2\,y + z = 6$, has the same direction vectors.

(b) Find a position vector: We can use any point that satisfies the equation $x + 2\,y + z = 6$. Since there is only one plane with a given set of direction vectors that passes through a given point in R^3, this will guarantee that we have the described the required plane. One choice of p is found by letting $y = 0$ and $z = 0$ and solving for x. This yields $p = [\,6\,,0\,,0\,]$.

[>

Solution Method 2: We derive the vector representation in two steps by using backsolve(..) to express the general solution of $x + 2\,y + z = 6$ and then decomposing the solution.

```
> eqn2 := x+2*y+z=6;
> M := genmatrix([eqn2],[x,y,z],flag);
> soln := backsolve(M,t);
```

[>

Now take a good look at the vector "soln" above. Notice that we can break it down into three parts, much as we did in Example 2A. This "decomposition" reveals the vectors p, v, and w.

$$\begin{bmatrix} 6-2\,t_2-t_1 \\ t_2 \\ t_1 \end{bmatrix} = \begin{bmatrix} 6 \\ 0 \\ 0 \end{bmatrix} + t_2 \begin{bmatrix} -2 \\ 1 \\ 0 \end{bmatrix} + t_1 \begin{bmatrix} -1 \\ 0 \\ 1 \end{bmatrix}$$

```
[ >
```

====================

As an alternative, here is a systematic Maple-based procedure that produces the decomposition:

Systematic Procedure for Decomposing the Solution Using Maple

Step 1. Set all free variables to 0, which gives you a position vector *p*.

```
[ > p := subs({t[1]=0,t[2]=0},evalm(soln));
```

Step 2. Subtract *p* from the general solution. This will leave you with the span of the direction vectors, $t_1\,v + t_2\,w$. This is the part of the solution that corresponds to the parallel plane through the origin.

```
[ > h := evalm(soln-p);
```

Step 3. In the solution set for the plane through the origin, set one free variable to 1 and the rest to 0. This gives one of the direction vectors. Repeat for each of the free variables.

```
[ > v := subs({t[1]=1,t[2]=0},evalm(h));
[ > w := subs({t[1]=0,t[2]=1},evalm(h));
```

Step 4. Check that $p + t_1\,v + t_2\,w$ equals the solution set "soln".

```
> evalm(p+t[1]*v+t[2]*w);
> evalm(soln);
[ >
```

Section 4: Decomposing the Solution of a Linear System.

In this section, we will decompose the solution set of an arbitrary linear system and not limit ourselves to systems in two and three unknowns. Our approach will draw on our experience with vector representations $p + t\,v$ of a line and $p + t\,v + s\,w$ of a plane. In fact, the decomposition method we will use in this section will be a straightforward extension of the decompositions in Section 3.

====================

Example 4A: Consider the following linear system consisting of three equations in four unknowns.

$$2\,x_1 - 6\,x_2 + 3\,x_3 - 2\,x_4 = -1$$
$$-x_1 + 3\,x_2 - 2\,x_3 = 4$$
$$3\,x_1 - 9\,x_2 + 4\,x_3 - 4\,x_4 = 2$$

Execute the next input region to compute the general solution of this system.

```
> eqn1 := 2*x[1]-6*x[2]+3*x[3]-2*x[4]=-1:
> eqn2 := -x[1]+3*x[2]-2*x[3]=4:
```

```
> eqn3 := 3*x[1]-9*x[2]+4*x[3]-4*x[4]=2:
> M := genmatrix([eqn1,eqn2,eqn3],[x[1],x[2],x[3],x[4]],flag);
> soln := backsolve(gausselim(M));
>
```

====================

Exercise 4.1: Decompose this general solution into the form $p + t_1\, v + t_2\, w$.

Student Workspace

Answer 4.1

Notice that the solution of this linear system has two free variables, t_1 and t_2, which is the same number of free variables we had in the case of the plane in Example 3A (Section 3). Even though the solution is a set of points in R^4, it will be useful to think of it as if it were also a "plane." In particular, in writing the solution in the form $p + t_1\, v + t_2\, w$, it is natural to refer to p as a *position vector* for the solution, and similarly we will refer to v and w as nonparallel *direction vectors* for the "parallel plane" through the origin. The vector p is also called a *particular solution* of the linear system.

```
>
```

Problems

Problem 1: Vector representation of a plane

Find a vector representation $p + t\, v + s\, w$ of the plane $x + 2\, y + 3\, z = 12$. Explain your work.

Student Workspace

Problem 2: Picture of a plane and associated vectors

Use plot3d(..), drawvec3d(..), and display3d(..) to produce a plot that includes all of the following: the graph of $x + 2\, y + 3\, z = 12$, the graph of the parallel plane through the origin ($x + 2\, y + 3\, z = 0$), the vectors p, v, w you found in Problem 1, and the normal vector n to the plane. Suggestion: You may find it difficult to orient the plot so it displays both planes and all four vectors in the way you want them to look. In that case, produce two separate plots (but no more than two). If necessary, remove some of the vectors from one of the plots. (Don't spend a lot of time, however, getting the pictures to look just right.)

Student Workspace

Problem 3: Vector representation of the intersection of two planes

(a) Below are the equations of two planes that intersect in a line. Solve the system of two equations, and express the solution set (i.e., the line of intersection) in vector form.
(b) How is the direction vector of the line related geometrically to the normal vectors of the two

planes?

```
> eqn1 := 2*x-y+3*z=6;
> eqn2 := x-4*z=8;
> solve({eqn1,eqn2});
>
```

Student Workspace

Problem 4: Picture of parallel parallelograms

Use polygonplot3d(..) to draw the following two parallelograms in R^3. The first parallelogram has vertices at [1, 2, 0], [0, 0, 0], [2, −3, 1], and the fourth vertex opposite [0, 0, 0]. The second parallelogram is parallel to the first one but translated by the vector $p = [3, 0, 2]$. Below is a sample of how polygonplot3d(..) is used. The argument in polygonplot3d(..) is a list of points; polygonplot3d(..) connects consecutive points by line segments and also connects the last point back to the first point. In the sample below, we connected three of the given points to form a triangle.

```
> polygonplot3d([[1,2,0],[0,0,0],[2,-3,1]]);
>
```

Student Workspace

Problem 5: Decompose the solution of a linear system I

Solve the following system of equations, and decompose the solution into a linear combination of vectors. Identify the position and direction vectors that generate the solution.

$$x_1 + x_2 - 2\,x_3 + x_4 = 3$$
$$2\,x_1 + 2\,x_2 - x_3 + 2\,x_4 = 3$$
$$3\,x_1 + 3\,x_2 + 2\,x_3 + 3\,x_4 = 1$$

To get started, execute the next input region to define the three equations and the corresponding augmented matrix.

```
> eq1 := x[1]+x[2]-2*x[3]+x[4]=3;
  eq2 := 2*x[1]+2*x[2]-x[3]+2*x[4]=3;
  eq3 := 3*x[1]+3*x[2]+2*x[3]+3*x[4]=1;
  A := genmatrix([eq1,eq2,eq3],[x[1],x[2],x[3],x[4]],flag);
>
```

Student Workspace

Problem 6: Decompose the solution of a linear system II

Solve the following system of equations, and decompose the solution into a linear combination of vectors. Identify the position and direction vectors that generate the solution.

$$x_1 + x_2 + 3\,x_4 + 3\,x_5 = 2$$
$$-2\,x_1 + 2\,x_2 - x_3 + 2\,x_4 + 3\,x_5 = 2$$
$$4\,x_1 + x_3 + 4\,x_4 + 3\,x_5 = 2$$
$$-7\,x_1 + 5\,x_2 - 3\,x_3 + 3\,x_4 + 6\,x_5 = 4$$

To get started, execute the next input region to define the four equations and the corresponding augmented matrix.

```
> eqn1 := x[1]+x[2]+3*x[4]+3*x[5]=2;
> eqn2 := -2*x[1]+2*x[2]-x[3]+2*x[4]+3*x[5]=2;
> eqn3 := 4*x[1]+x[3]+4*x[4]+3*x[5]=2;
> eqn4 := -7*x[1]+5*x[2]-3*x[3]+3*x[4]+6*x[5]=4;
> M := genmatrix([eqn1,eqn2,eqn3,eqn4],
    [x[1],x[2],x[3],x[4],x[5]],flag);
>
```

Student Workspace

Problem 7: Nonhomogeneous and homogeneous systems

Here is a system of linear homogeneous equations and two right-side column vectors b and c:

$$4\,x_1 + x_2 + 2\,x_3 - 3\,x_4 - 2\,x_5 = 0$$
$$-2\,x_1 - x_2 - x_3 + 4\,x_4 + x_5 = 0$$
$$4\,x_2 - 3\,x_3 + 2\,x_4 + 2\,x_5 = 0$$

$$b = \begin{bmatrix} 13 \\ -7 \\ 1 \end{bmatrix} \quad c = \begin{bmatrix} 6 \\ -7 \\ 6 \end{bmatrix}$$

(a) Solve the homogeneous system and decompose the solution. Do the same for each of the two nonhomogenous systems that have the same coefficient matrix and whose right-sides are b and c, respectively. (The commands in the input region below will help you get started.)

(b) How are the three solution sets you found in (a) related to one another? That is, describe in words the similarities and differences among the three solution sets. In your description, use the geometric language we employed in Section 3.

```
> A := matrix([[4,1,2,-3,-2],[-2,-1,-1,4,1],[0,4,-3,2,2]]);
> aug1 := augment(A,colvector([0,0,0]));
> aug2 := augment(A,colvector([13,-7,1]));
> aug3 := augment(A,colvector([6,-7,6]));
>
```

Student Workspace

Pencil and Paper Practice Questions

Note: Maple is not required for any of the problems in this section. Do all of these by hand.

1. Express the plane $3\,x - 4\,y + 5\,z = 0$ in the parametric form $t\,v + s\,w$.

2. Let $u = \begin{bmatrix} 2 \\ -1 \end{bmatrix}$, $v = \begin{bmatrix} -1 \\ 3 \end{bmatrix}$, and $p = \begin{bmatrix} 4 \\ 2 \end{bmatrix}$. Draw the following set of points, and describe it in words:

$$\{p + a\, u + b\, v \mid a \text{ is between -1 and 2, and } b \text{ is between 0 and 1}\}$$

3. Below is a solution set of a linear system. Decompose the solution into a linear combination of vectors. Identify the position and direction vectors that generate the solution.

$$[3\, t_1 - 4\, t_2 + t_3 - 6,\ 2\, t_1 - 2\, t_3 + 1,\ -t_2 + 3\, t_3,\ t_3,\ -5\, t_2 + 9,\ t_1,\ t_2]$$

Linear Algebra Modules Project
Chapter 2, Module 4

Linear Independence of Vectors

Purpose of this module

The purpose of this module is to introduce the concepts of linearly dependent and linearly independent vectors.

Prerequisites

Linear combinations and span of a set of vectors. Solution of a system of linear equations.

Commands used in this module

```
> restart: with(linalg): with(plots): with(lamp):
>
```

Tutorial

Section 1. Linear Dependence of Vectors: A Geometric Introduction

In Module 2 we observed that any two nonparallel vectors u and v in R^2 span all of 2-space. In other words, every vector in R^2 can be expressed as a linear combination of the vectors u and v. If u and v are nonparallel vectors in R^3, their span is a plane in R^3. So two vectors cannot span all of R^3; we will need at least three vectors in R^3 to span the entire space. Of course, one set of three vectors that spans all of R^3 consists of the standard unit vectors i, j, k.

The question we ask is this: Will any three nonparallel vectors in R^3 span all of 3-space ? Before reading on, take a moment to think about this question and form a preliminary answer.

====================

Example 1A: Consider the set of three vectors $\{ u, v, p \}$ defined below. Note that the vectors are nonparallel; that is, they all point in different directions. Does Span$\{ u, v, p \} = R^3$?

```
> u := colvector([3,4,1]);
  v := colvector([5,1,2]);
  p := colvector([1,7,0]);
```

Let's start by looking at a picture of the three vectors:

```
> drawvec3d(u,v,p);
>
```

By turning the picture in different directions, you should be able to convince yourself that the three vectors lie in a single plane. (Try using the viewing angles $\Theta = 139$ degrees and $\Phi = 110$ degrees, which gives you a side view of the plane containing the three vectors.) Of course, no such picture can be conclusive.

To determine conclusively that the three vectors lie in a single plane, we need an algebraic method for testing them. Note that if all three vectors do lie in a single plane and no two are parallel, then each of the vectors must be in the span of the other two. Therefore each vector can be expressed as a linear combination of the other two.

So let's attempt to find a solution to the vector equation $p = a\,u + b\,v$. We can solve this equation by using the corresponding augmented matrix:

```
[ > N := augment(u,v,p);
[ > backsolve(gausselim(N));
[ >
```

A solution exists! In fact, we have shown that: $p = 2\,u - v$. This equation relating *u*, *v*, and *p* is called a *linear dependence relation*. Sometimes we will write this equation with all the given vectors on one side of the equation, like this: $0 = 2\,u - v - p$, where the 0 on the left side of the equation is the zero vector in R^3.

====================

In the next Exercise you will see how the existence of this linear dependence relation leads us to the conclusion that the Span{ *u*, *v*, *p* } is only a plane and not all of R^3.

```
[ >
```

Exercise 1.1:

(a) (By hand) Prove that Span{ *u*, *v*, *p* } = Span{ *u*, *v* } by demonstrating that every linear combination of the three vectors *u*, *v*, *p* can be rewritten as a linear combination of just the two vectors *u* and *v*.

(b) Illustrate the result from part (a) by showing how the vector $4\,u + 5\,v + 3\,p$ can be rewritten as a linear combination of the vectors *u* and *v*. Check your work by direct calculation.

⊞ Student Workspace

⊞ Answer 1.1

====================

Example 1B: Consider the set of three vectors { *u*, *v*, *w* } defined below. The first two vectors are the same as in Example 1A, but we have replaced *p* by a new vector *w*. In particular, we have chosen *w* so that it does not lie in the same plane as *u* and *v*.

```
> u := colvector([3,4,1]);
  v := colvector([5,1,2]);
  w := colvector([2,2,4]);
```

The picture below illustrates the fact that the vector *w* (pointing upward) lies outside the plane spanned by the vectors *u* and *v*.

```
[ > drawvec3d(u,v,w);
[ >
```

====================

Exercise 1.2:
(a) Prove algebraically that w is not in the plane spanned by u and v.
(b) Explain why the fact that w is not in the plane spanned by u and v implies that: (i) u is not in the plane spanned by the vectors v and w and (ii) v is not in the plane spanned by the vectors u and w.
(c) Explain why no linear dependence relation can exist between the vectors u, v, w. Hint: Show that such a linear dependence relation would lead to a contradiction.

Student Workspace

Answer 1.2

[>

In Exercise 1.2 you proved that no linear dependence relation exists among the vectors u, v, w. When a set of vectors has this property, we call it a *linearly independent* set. In the next Exercise, you will prove that the linearly independent set of vectors $\{u, v, w\}$ spans all of 3-space.

Exercise 1.3: Show that Span$\{u, v, w\} = R^3$. Hint: Take a general vector in R^3, say $q = [x, y, z]$, and show that q can be expressed as a linear combination of u, v, w by showing that the vector equation $a\,u + b\,v + c\,w = q$ always has a solution.

Student Workspace

Answer 1.3

We began this section by asking whether any set of three nonparallel vectors in R^3 will span 3-space. Example 1A demonstrated that it is not enough for the vectors to be merely nonparallel. Example 1B provided an example of a linearly independent set that spans all of 3-space. We will soon see that linearly independence is, in fact, a necessary and sufficient condition for three vectors to span 3-space. Before we can pursue this topic further, we need to take a closer look at the definitions of linear dependence and independence. The next section presents a systematic algebraic approach to these fundamental concepts.

[>

Section 2. Linear Dependence of Vectors: An Algebraic Approach

Sometimes a given set of vectors satisfies a linear dependence relation and sometimes it does not. When the set of vectors does satisfy a linear dependence relation, we say that the set of vectors is *linearly dependent*; if the set of vectors does not satisfy any linear dependence relation, we say that the set of vectors is *linearly independent*. Here are the formal definitions:

Definition: A set of vectors $\{v_1, v_2, \ldots, v_k\}$ in R^n is ***linearly dependent*** if the vectors satisfy a ***linear dependence relation***

$$c_1\,v_1 + c_2\,v_2 + \ldots + c_k\,v_k = 0$$

where at least one of the weights $c_1, c_2, ..., c_k$ is not zero. The set of vectors is ***linearly independent*** if it is not linearly dependent, that is, if there is no linear dependence relation satisfied by the vectors.

[>

Let's take a close look at a set of vectors that is linearly dependent.

====================

Example 2A: Consider the set of three vectors { *u*, *v*, *w* } defined below.

```
> u := colvector([2,3]);
  v := colvector([4,-1]);
  w := colvector([14,7]);
```

[>

This set of three vectors satisfies the linear dependence relation $3\,u + 2\,v - w = 0$. (Check this by hand.) Therefore { *u*, *v*, *w* } is a linearly dependent set of vectors. Notice also that each of the three vectors *u*, *v*, *w* is a linear combination of the other two:

$$w = 3\,u + 2\,v, \quad v = \frac{w}{2} - \frac{3\,u}{2}, \quad u = \frac{w}{3} - \frac{2\,v}{3}$$

====================

====================

Example 2B: Consider the set of vectors { *u*, *v*, *w* } defined below.

```
> u := colvector([2,3]);
  v := colvector([4,6]);
  w := colvector([14,7]);
```

[>

This set of vectors satisfies the linear dependence relation $2\,u - v + 0\,w = 0$. (Check this by hand.) Therefore { *u*, *v*, *w* } is a linearly dependent set of vectors. Unlike the previous example, however, it is not true that each of the three vectors is a linear combination of the other two. This time, *u* is a linear combination of *v* and *w*, and *v* is a linear combination of *u* and *w*, but *w* is not a linear combination of *u* and *v*:

$$u = \frac{v}{2} + 0\,w, \quad v = 2\,u + 0\,w$$

====================

[>

Let's look again at the definition of linear dependence. Note that the equation $c_1\,v_1 + c_2\,v_2 + ... + c_k\,v_k = 0$ can always be solved for $c_1, c_2, ..., c_k$ simply by choosing all the weights to be zero. However, the equation $0\,c_1 + 0\,c_2 + ... + 0\,c_k = 0$ does <u>not</u> constitute a linear dependence relation. We refer to the solution in which all the weights are zero as the "trivial" solution. So here is another way to express the statement that the set of vectors $\{ v_1, v_2, ..., v_k \}$ is linearly dependent:

- The set $\{ v_1, v_2, ..., v_k \}$ is linearly dependent if and only if the equation $c_1 v_1 + c_2 v_2 + ... + c_k v_k = 0$ has a nontrivial solution (i.e., a solution in which at least one of the weights is nonzero).

====================

Example 2C: Let's find out whether the set of vectors $\{ u, v, w \}$ defined below is linearly dependent or independent.

```
> u := colvector([3,4,1]);
  v := colvector([5,1,2]);
  w := colvector([2,2,4]);
```

Solution: We need to find out if there exist weights c_1, c_2, c_3 (where at least one is not zero) such that

$$c_1 u + c_2 v + c_3 w = 0$$

We can solve this equation by using the corresponding augmented matrix:

```
> N := augment(u,v,w,colvector([0,0,0]));
> backsolve(gausselim(N));
>
```

Since the only solution to the vector equation is the trivial solution in which the weights c_1, c_2, c_3 are all zero, there is no linear dependence relation satisfied by $\{ u, v, w \}$, and hence the set $\{ u, v, w \}$ is linearly independent.

====================

Let's look at these three examples more closely to see what general conclusions we can infer. A set of vectors $\{ v_1, v_2, ..., v_k \}$ is defined to be linearly dependent if there is a linear dependence relation $c_1 v_1 + c_2 v_2 + ... + c_k v_k = 0$, where at least one of the weights $c_1, c_2, ..., c_k$ is not zero. Since at least one of the weights is not zero in such a linear dependence relation, we can solve for at least one of the vectors, thus writing that vector as a linear combination of the other vectors. (See Examples 2A and 2B.) However, as we saw in Example 2B, some of the weights can be zero and we still say that the entire set of vectors is linearly dependent. So we have another way to express the statement that the set of vectors $\{ v_1, v_2, ..., v_k \}$ is linearly dependent:

- The set of vectors $\{ v_1, v_2, ..., v_k \}$ is linearly dependent if and only if at least one of the vectors can be written as a linear combination of the others.

```
>
```

Exercise 2.1: Consider the set of four vectors $\{ v_1, v_2, v_3, v_4 \}$ in R^5 defined below.

(a) Show that this set of vectors is linearly dependent, and find a linear dependence relation.

(b) Solve for one of the vectors as a linear combination of the others.

(c) Now consider the set consisting of only the three vectors that remain after you remove the vector you solved for in (b). Is this set linearly dependent or independent?

```
> v1 := colvector([4,3,-1,2,1]);
  v2 := colvector([-4,7,-9,0,3]);
  v3 := colvector([2,3,-2,0,1]);
  v4 := colvector([0,2,-2,3,1]);
```

[>

Student Workspace

Answer 2.1

==================

Example 2D -- The Special Case of a Set of Two Vectors: If a set consists of only two vectors, say u and v, then solving the vector equation $a\,u + bv = 0$ is equivalent to determining whether one of the vectors is a multiple of the other. For example, the equation $u = 3\,v$ is equivalent to the linear dependence relation $u - 3\,v = 0$. In general, the linear dependence relation $a\,u + bv = 0$ is equivalent to the equation $u = \left(-\frac{b}{a}\right)v$, provided $a \neq 0$.

==================

Exercise 2.2: (By hand) For each pair of vectors listed below, determine whether the set consisting of the two vectors is linearly dependent or independent. If it is linearly dependent, find a linear dependence relation.

[>

(a) $v_1 = \begin{bmatrix} 2 \\ 3 \\ 5 \end{bmatrix}$ and $v_2 = \begin{bmatrix} 8 \\ 12 \\ 20 \end{bmatrix}$

(b) $v_1 = \begin{bmatrix} 2 \\ 3 \\ 5 \end{bmatrix}$ and $v_2 = \begin{bmatrix} 8 \\ 12 \\ 25 \end{bmatrix}$

(c) $v_1 = \begin{bmatrix} 2 \\ 3 \\ 5 \end{bmatrix}$ and $v_2 = \begin{bmatrix} 0 \\ 0 \\ 0 \end{bmatrix}$

[>

Student Workspace

Answer 2.2

[>

Section 3. More Examples of Linear Independence and Dependence

If we have a set of vectors that is linearly dependent and we add a vector to the set or remove a vector from the set, can we predict whether the new set is linearly dependent or independent? What if the original set is linearly independent? We examine these questions in this section.

[>

Suppose { v_1, ... ,v_k } is a linearly dependent set of vectors in R^n. Then $\{v_1,...,v_k\}$ must satisfy a linear dependence relation

$$c_1 v_1 + c_2 v_2 + ... + c_k v_k = 0$$

where at least one of the weights $c_1, c_2, ..., c_k$ is nonzero. So if w is another vector in R^n then, since $0\, w = 0$, we have the linear dependence relation

$$c_1 v_1 + c_2 v_2 + ... + c_k v_k + 0\, w = 0$$

with the same coefficients $c_1, c_2, ... c_k$ as above and the additional coefficient 0. We have thus proved:

- If $\{ v_1, ... , v_k \}$ is a linearly dependent set of k vectors in R^n and w is any vector in R^n, then the set of $k + 1$ vectors $\{ v_1, ... , v_k , w \}$ must also be a linearly dependent set.

```
>
```

On the other hand, adding a vector to a set of linearly independent vectors may or may not result in a linearly independent set.

Exercise 3.1: The set of vectors $\{ u, v \}$ below is linearly independent. Find vectors p and q such that $\{ u, v, p \}$ is linearly independent and $\{ u, v, q \}$ is linearly dependent. Explain your answer in geometric terms.

```
> u := colvector([1,-1,0]);
  v := colvector([0,1,-1]);
>
```

Student Workspace

Answer 3.1

Similarly, removing a vector from a linearly dependent set may or may not result in a linearly dependent set.

Exercise 3.2:

(a) Consider the four vectors in R^3 shown below. This set of four vectors, $\{ u, v, x, y \}$, is linearly dependent. Which vector could you remove from this set so the remaining three vectors still form a linearly dependent set? Is there more than one choice? Which vector could you remove from this set so that the remaining three vectors form a linearly independent set? Is there more than one choice?

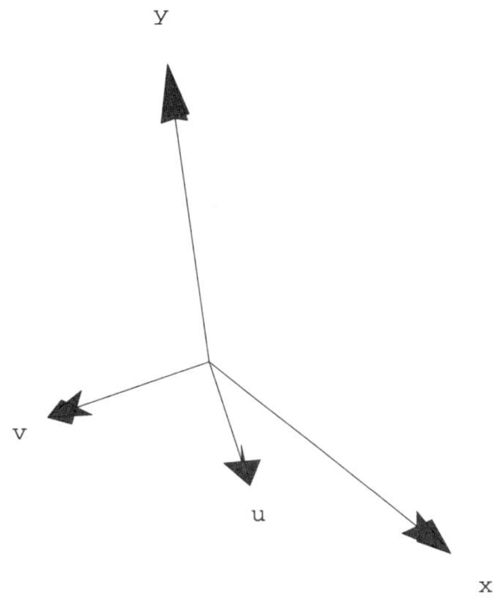

[>

(b) Consider the set of four vectors in R^3 shown below. This set is also linearly dependent. Can you remove a vector from this set so the remaining three form a linearly dependent set? Which vector or vectors could you remove from this set so that the remaining three vectors form a linearly independent set?

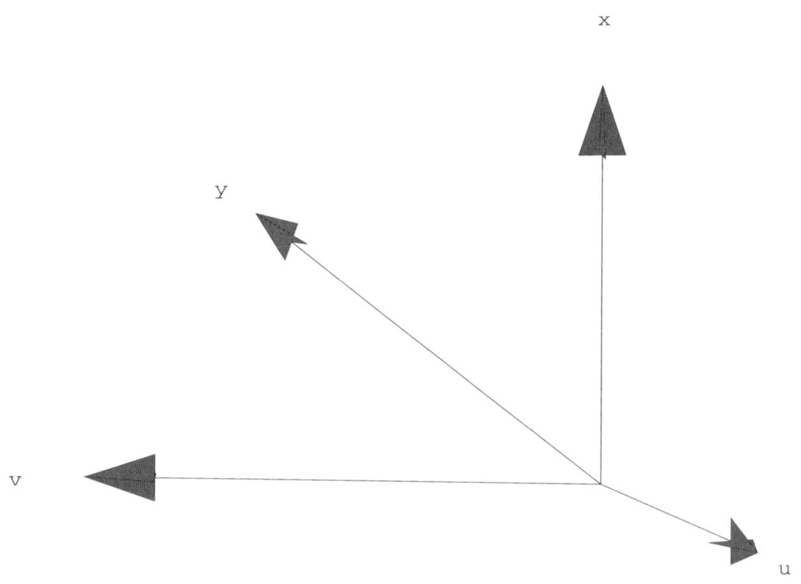

[>

Student Workspace

Answer 3.2

[>

Linear Independence of Solutions Produced by Decomposition

==================

Example 3A: In Module 3, we saw how to decompose the solution of a linear system. The general form we obtained from such a decomposition was p + Span$\{v_1, v_2, ..., v_k\}$, where p is some particular solution of the given linear system, and Span$\{v_1, v_2, ..., v_k\}$ is the set of solutions of the corresponding homogeneous linear system. The result of any decomposition has a geometric interpretation: We can think of the set of solutions of the homogeneous system as analogous to a plane through the origin and the set of solutions of the given system as the translation of this "plane" by the vector p.

In this example, we will use the method of decomposition to find a linearly independent set of vectors whose span is the set of solutions of a given homogeneous linear system:

```
> x := 'x':
> eqn1 := 2*x[1]-6*x[2]+3*x[3]-2*x[4]=0:
> eqn2 := -x[1]+3*x[2]-2*x[3]=0:
```

```
> eqn3 := 3*x[1]-9*x[2]+4*x[3]-4*x[4]=0:
> M := genmatrix([eqn1,eqn2,eqn3],[x[1],x[2],x[3],x[4]],flag);
> backsolve(gausselim(M));
>
```

When we decompose this set of solutions, we see that it is the span of $v_1 = [4, 0, -2, 1]$ and $v_2 = [3, 1, 0, 0]$. Note that $\{v_1, v_2\}$ is a linearly independent set of vectors.

==================

That the method of decomposition in this example produced a linearly independent spanning set was not an accident. You will see in Problem 9 that:

- When the method of decomposition is applied to the solution set of a homogeneous linear system, the vectors produced by the decomposition are always linearly independent.

```
>
```

Problems

Problem 1: Linear independence and linear combinations

(a) Show that the set of vectors $\{v_1, v_2, v_3, v_4\}$ below is linearly dependent, and find one linear dependence relation for this set of vectors.
(b) Which of the four vectors is a linear combination of the other three? Give all possible answers.

```
> v1 := colvector([3,-1,2,0]);
  v2 := colvector([-5,0,-3,2]);
  v3 := colvector([1,-2,1,2]);
  v4 := colvector([0,4,-2,3]);
>
```

Student Workspace

Problem 2: Does pairwise independence imply independence?

True or false: "If v_1, v_2, v_3 are vectors in R^3 and the sets $\{v_1, v_2\}$, $\{v_2, v_3\}$ and $\{v_1, v_3\}$ are each linearly independent, then the set $\{v_1, v_2, v_3\}$ must also be linearly independent." If this statement is true, explain why; if it is not, give a counterexample.

Student Workspace

Problem 3: Any four vectors in R^3 are linearly dependent

(a) Consider the four vectors $\{v_1, v_2, v_3, v_4\}$ in R^3 defined below. Demonstrate that these vectors are linearly dependent.
(b) Explain why every set of four vectors in R^3 must be linearly dependent.

```
> v1 := colvector([1,3,5]):
> v2 := colvector([2,4,1]):
  v3 := colvector([-3,0,3]):
> v4 := colvector([3,1,4]):
>
```

Student Workspace

Problem 4: Find the linearly independent subsets

Confirm that the set of vectors $\{ v_1, v_2, v_3, v_4, v_5 \}$ defined below is linearly dependent. Which of the subsets $\{ v_1, v_2 \}$, $\{ v_1, v_2, v_3 \}$, $\{ v_1, v_2, v_5 \}$, $\{ v_1, v_2, v_3, v_4 \}$ is linearly independent? In answering this question, do as few computations as possible; give explanations instead. For example, once you have answered this question for one subset, you may be able immediately to answer the question for another subset.

```
> v1 := colvector([-4, -2, -1, 4, 7]):
  v2 := colvector([7, 1, -2, -6, -8]):
  v3 := colvector([-3, 3, 0, 2, 4]):
  v4 := colvector([-6, -6, -1, -2, 3]):
  v5 := colvector([1, 3, 4, -2, -6]):
> M := augment(v1,v2,v3,v4,v5,colvector([0,0,0,0,0]));
>
```

Student Workspace

Problem 5: Linear dependence and the zero vector

True or false: "If $\{v_1, v_2, ..., v_k\}$ is a set of vectors in R^n that includes the zero vector, the set is linearly dependent." If this statement is true, explain why; if it is not, give a counterexample.

Student Workspace

Problem 6: Linearly independent subsets of a linearly dependent set

Find a set of four vectors in R^4 that is linearly dependent but every subset of which is linearly independent (except, of course, for the set itself). For example, if the set of four vectors is $\{ v_1, v_2, v_3, v_4 \}$, then subsets such as $\{ v_1, v_2, v_3 \}$, $\{ v_2, v_3, v_4 \}$, $\{ v_1, v_2 \}$, etc. are all independent sets.

Student Workspace

Problem 7: Linearly dependent subset of a linearly independent set?

Let v_1 and v_2 be two vectors in R^4 defined below. If possible, find two additional vectors v_3 and v_4 such that the set $\{ v_1, v_2, v_3, v_4 \}$ is linearly independent but the set $\{ v_1, v_2, v_3 \}$ is linearly dependent. If this is not possible, explain why not.

```
> v1 := colvector([2,3,4,5]);
  v2 := colvector([1,3,5,4]);
>
```

Student Workspace

Problem 8: Concrete application (part 2)

This problem follows up on Problem 5 of Module 2 of Chapter 2. Below is the table that describes the composition of the three basic mixtures of concrete, S, A, L. Consider two custom mixtures, X and Y, below. (All five vectors have been entered below.)

```
>
```

	Super-Strong *Type S*	*All-Purpose* *Type A*	*Long-life* *Type L*
cement	20	18	12
water	10	10	10
sand	20	25	15
gravel	10	5	15
fly ash	0	2	8

	Type X	*Type Y*
cement	12	15
water	12	10
sand	12	20
gravel	12	10
fly ash	12	5

```
> S := colvector([20,10,20,10,0]):
> A := colvector([18,10,25,5,2]):
> L := colvector([12,10,15,15,8]):
> X := colvector([12,12,12,12,12]):
> Y := colvector([15,10,20,10,5]):
>
```

(a) Show that {S, A, L} is a linearly independent set of vectors. What practical advantage does that have?

(b) Show that we can make the custom mix Y but not the custom mix X from the three basic mixes S, A, L. What does this say about the linear independence or dependence of the sets {S, A, L, X} and {S, A, L, Y}?

(c) Explain why any combination of S, A, L and Y can be also achieved by a combination of just S, A and L. For example, show how to make the custom mix $3\,S + 4\,A + 2\,L + 3\,Y$ using only S, A and L.

(d) Define a fifth basic mix Z to add to {S, A, L, X} such that any custom mixture can be expressed as a linear combination of the set of mixes { S, A, L, X, Z }.

(e) Why will there still be mixes that cannot be physically produced from this set of five basic mixes? (Hint: Consider the signs of the scalar weights.) Give an example of such a mix.

Student Workspace

Problem 9: Linear independence and decomposition

(a) Use decomposition to express the set of solutions of the homogeneous linear system below in the form Span$\{v_1, v_2, ..., v_k\}$. By hand, check that the set of vectors $\{v_1, v_2, ..., v_k\}$ is linearly independent, using the following quick method: When you write out the equation of the form $c_1 v_1 + c_2 v_2 + ... + c_k v_k = 0$ and combine the terms on the left, note that this equation is exactly the equation you would get if you were to set the result of the backsolve(..) command below equal to 0. Then look at the components that correspond to the free variables in the solution of the linear system.

(b) (Challenge) Explain why the method of decomposition, when applied to the solution set of a homogeneous linear system, always yields a linearly independent set of vectors whose span is the set of solutions. Hint: Use the quick method in (a) that helped you check linear independence by hand.

```
> eqn1 := -2*x[1]+3*x[2]-x[3]+x[4]+4*x[5]+x[6]+4*x[7] = 0;
> eqn2 := x[1]-3*x[2]-2*x[4]-3*x[5]+x[7] = 0;
> eqn3 := -5*x[1]+x[2]+3*x[3]-4*x[4]-2*x[5]-3*x[6] = 0;
> M := genmatrix([eqn1,eqn2,eqn3],
    [x[1],x[2],x[3],x[4],x[5],x[6],x[7]],flag);
> backsolve(gausselim(M));
>
```

Student Workspace

Pencil and Paper Practice Questions

Note: Maple is not required for any of the problems in this section. Do all of these by hand.

1. (a) Find a vector v_3 so that $\{ v_1, v_2, v_3 \}$ is linearly independent, where v_1, v_2 are given below.
(b) Show that $\{ v_1, v_2, v_3 \}$ is linearly independent, directly from the definition of linear independence.

$$v_1 = \begin{bmatrix} 2 \\ 0 \\ -1 \end{bmatrix} \text{ and } v_2 = \begin{bmatrix} -2 \\ 3 \\ 0 \end{bmatrix}$$

2. (a) Find a linear dependence relation for the following vectors. (The relation is not a hard one to spot; you can find it without a computation.)

$$u = \begin{bmatrix} 2 \\ 1 \\ -1 \\ 0 \end{bmatrix}, \quad v = \begin{bmatrix} -1 \\ -3 \\ 0 \\ 1 \end{bmatrix}, \quad w = \begin{bmatrix} 0 \\ -5 \\ -1 \\ 2 \end{bmatrix}$$

(b) If x is any other vector in R^4, show that $\{ u, v, w, x \}$ is linearly dependent too; write down a linear dependence relation for these four vectors.

Linear Algebra Modules Project
Chapter 3, Module 1

Product of a Matrix and a Vector

Purpose of this module

The purpose of this module is to acquaint you with the definition of matrix-vector multiplication, its properties, and its uses.

Prerequisites

Algebra and geometry of vectors, linear independence, span.

Commands used in this module

```
> restart:  with(linalg): with(lamp):
>
```

Tutorial

Section 1: Definition of the Matrix-Vector Product $A\,x$

Definition: Suppose A is a matrix with m rows and n columns, and x is a vector with n components. The ***matrix-vector product*** $A\,x$ is defined to be the linear combination of the columns of A in which the weights are the components of x.

We can express this definition in symbols: Let $A_1, A_2, \ldots, A_n$ denote the columns of the matrix A, and $x_1, x_2, \ldots, x_n$ denote the components of the vector x. Then

$$A\,x = [A_1 \quad A_2 \quad \ldots \quad A_n]\begin{bmatrix} x_1 \\ \cdot \\ \cdot \\ x_n \end{bmatrix} = x_1\,A_1 + x_2\,A_2 + \ldots + x_n\,A_n$$

```
>
```

====================

Example 1A: Find the product $A\,x$ for the matrix $A = \begin{bmatrix} 1 & 3 & -2 \\ 2 & -1 & 4 \end{bmatrix}$ and vector $x = \begin{bmatrix} 5 \\ 3 \\ 2 \end{bmatrix}$.

Solution: The columns of A are $A_1 = \begin{bmatrix} 1 \\ 2 \end{bmatrix}$, $A_2 = \begin{bmatrix} 3 \\ -1 \end{bmatrix}$, $A_3 = \begin{bmatrix} -2 \\ 4 \end{bmatrix}$, and the components of x are $x_1 = 5,\ x_2 = 3,\ x_3 = 2$. Using the components of x as weights and the columns of A as vectors, we calculate the product $A\,x$ as the linear combination:

$$5\begin{bmatrix} 1 \\ 2 \end{bmatrix} + 3\begin{bmatrix} 3 \\ -1 \end{bmatrix} + 2\begin{bmatrix} -2 \\ 4 \end{bmatrix} = \begin{bmatrix} 10 \\ 15 \end{bmatrix}$$

====================

The Maple commands to compute this product are as follows:

```
> A := matrix([[1,3,-2],[2,-1,4]]);
  x := colvector([5,3,2]);
> evalm(A&*x);
>
```

Note that you must use evalm(..) to force Maple to evaluate the product, and that Maple's multiplication symbol for matrix multiplication is &* instead of simply *. Find out what happens if you eliminate evalm(..) or & from the above command.

Exercise 1.1: Compute the matrix-vector product $M\,c$ using the matrix M and vector c below. First do this by hand to practice the definition above. Then check your answer by having Maple calculate the product.

```
> M := matrix([[3,-2],[-4,2],[8,5]]);
> c := colvector([1,2]);
>
```

Student Workspace

Answer 1.1

Exercise 1.2: Below we define vectors u, v, w and compute the linear combination $p = 2\,u + 11\,v + 7\,w$. Find a matrix A and vector x such that $p = A\,x$.

```
> u := colvector([2,3,8]);
  v := colvector([5,-6,0]);
  w := colvector([7,4,-2]);
> p := evalm(2*u+11*v+7*w);
>
```

Student Workspace

Answer 1.2

Exercise 1.3:

(a) If the vector x is simple enough, we can calculate $A\,x$ by inspection. Vectors that have 1's and

0's are especially easy to work with. By inspection, multiply the matrix A below by each of the vectors u, v, w, z below. (Just describe each linear combination in words.)

$$A = \begin{bmatrix} a & b & c \\ d & e & f \end{bmatrix}, \quad u = \begin{bmatrix} 1 \\ 0 \\ 0 \end{bmatrix} \quad v = \begin{bmatrix} 0 \\ 0 \\ 1 \end{bmatrix} \quad w = \begin{bmatrix} 0 \\ 1 \\ 1 \end{bmatrix} \quad z = \begin{bmatrix} 0 \\ 0 \\ 2 \end{bmatrix}$$

(b) Suppose N is a 8 by 5 matrix and you want to add the odd columns together to form a single 8 by 1 column vector. Describe how you could accomplish this using a matrix-vector product.

[>

Student Workspace

Answer 1.3

[>

Dot Product Method for Calculating $A\,x$

Let's take another look at the matrix-vector product that we calculated in Example 1A. Notice that the first component of the product $A\,x$ is simply the dot product of the first row of A and the vector x, and similarly for the second component:

$$5\begin{bmatrix} 1 \\ 2 \end{bmatrix} + 3\begin{bmatrix} 3 \\ -1 \end{bmatrix} + 2\begin{bmatrix} -2 \\ 4 \end{bmatrix} = \begin{bmatrix} 5(1) + 3(3) + 2(-2) \\ 5(2) + 3(-1) + 2(4) \end{bmatrix} = \begin{bmatrix} 10 \\ 15 \end{bmatrix}$$

So we now have a second way to compute the product $v = A\,x$: The ith component, v_i, of v is the dot product of the ith row of A with the vector x.

[>

Exercise 1.4: (By hand) Use the dot product method to calculate the following matrix products:

$$\text{(a)} \begin{bmatrix} 1 & 2 \\ 4 & 0 \\ -3 & 5 \end{bmatrix} \begin{bmatrix} 7 \\ 3 \end{bmatrix} \qquad \text{(b)} \begin{bmatrix} 5 & -2 & 3 \\ 7 & 10 & 4 \end{bmatrix} \begin{bmatrix} x_1 \\ x_2 \\ x_3 \end{bmatrix}$$

Student Workspace

Answer 1.4

[>

Section 2: Vector Equations, Matrix Equations and Linear Systems

We begin this section by reconsidering a problem that we first solved in Chapter 2. We will briefly repeat our earlier solution and then introduce a new way of looking at this familiar problem by using matrix-vector products.

====================

Example 2A: For the vectors u, v and b below, express b as a linear combination of u and v.

```
> u := colvector([2,3,4]);
  v := colvector([-4,2,1]);
> b := colvector([-40,4,-8]);
>
```

Solution method 1: To express b as a linear combination of u and v requires us to find weights x_1 and x_2 that satisfy the vector equation $x_1\, u + x_2\, v = b$:

$$x_1 \begin{bmatrix} 2 \\ 3 \\ 4 \end{bmatrix} + x_2 \begin{bmatrix} -4 \\ 2 \\ 1 \end{bmatrix} = \begin{bmatrix} -40 \\ 4 \\ -8 \end{bmatrix} \qquad [1]$$

Recall that in Chapter 2, Module 2, we solved a vector equation by writing it as a linear system:

$$\begin{aligned} 2\,x_1 - 4\,x_2 &= -40 \\ 3\,x_1 + 2\,x_2 &= 4 \\ 4\,x_1 + x_2 &= -8 \end{aligned} \qquad [2]$$

This linear system can then be solved by forming the augmented matrix of this linear system:

```
> M := augment(u,v,b);
  backsolve(gausselim(M));
>
```

So we have found that $b = -4\,u + 8\,v$.

Solution method 2: The left side of the vector equation [1] can be expressed as a matrix-vector product:

$$x_1 \begin{bmatrix} 2 \\ 3 \\ 4 \end{bmatrix} + x_2 \begin{bmatrix} -4 \\ 2 \\ 1 \end{bmatrix} = \begin{bmatrix} 2 & -4 \\ 3 & 2 \\ 4 & 1 \end{bmatrix} \begin{bmatrix} x_1 \\ x_2 \end{bmatrix}$$

Therefore equation [1] can be written in the form:

$$\begin{bmatrix} 2 & -4 \\ 3 & 2 \\ 4 & 1 \end{bmatrix} \begin{bmatrix} x_1 \\ x_2 \end{bmatrix} = \begin{bmatrix} -40 \\ 4 \\ -8 \end{bmatrix} \qquad [3]$$

Equation [3] can be written more succinctly yet as $A\,x = b$, where $A = \begin{bmatrix} 2 & -4 \\ 3 & 2 \\ 4 & 1 \end{bmatrix}$ and $x = \begin{bmatrix} x_1 \\ x_2 \end{bmatrix}$.

Thus, to solve the given problem, we simply need to solve the matrix equation $A\,x = b$ for the unknown vector x, which is accomplished by using the command matsolve(A,b):

```
[ > A := augment(u,v);
[ > matsolve(A,b);
[ >
```

Maple Note: The command matsolve(A,b) is equivalent to backsolve(gausselim(augment(A,b))). However, there is no need to think of matsolve(..) in such a complicated way; it simply finds all solutions x to the equation $A\,x = b$.

====================

In the next exercise, we ask you to practice moving between the three equivalent problems: a vector equation [1] , a linear system [2], and a matrix equation [3].

Exercise 2.1: Consider the vector equation $x_1\,u + x_2\,v = b$, where u, v, b are defined below:

```
[ > u := colvector([5,3]);
    v := colvector([3,2]);
    b := colvector([7,2]);
[ >
```

(a) Write out the corresponding linear system.

(b) Write out the corresponding matrix equation $A\,x = b$, and solve this equation using the matsolve(..) command.

(c) Check that the solutions that you found in (b) solve the vector equation and the linear system.

⊞ Student Workspace

⊞ Answer 2.1

The following theorem gives the general statement of the equivalence of vector equations, matrix equations and linear systems:

Theorem 1: Suppose A is a m by n matrix with columns A_1 , A_2 , ... , A_n and b is a vector in R^m. Then

- the vector equation $x_1\,A_1 + x_2\,A_2 + \dots + x_n\,A_n = b$,
- the matrix equation $A\,x = b$, and
- the linear system with coefficient matrix A and right-side vector b

all have the same set of solutions.

```
[ >
```

In the remainder of this section, we use matrix-vector products to review and deepen our understanding of the solution set of a linear system and the fundamental concepts of linear independence and span introduced in Chapter 2. First, however, we need some rules of algebra.

Rules of Algebra for Matrix-Vector Products

Two simple rules are valid for matrix-vector products:

$$A\,(x+y) = A\,x + A\,y \qquad \text{(distributive law)}$$
$$A\,(c\,x) = c\,(A\,x) \qquad \text{(scalars factor out)}$$

where A is any m by n matrix, x and y are each n by 1 column vectors, and c is a any real number.

====================

Example 2B: Use Maple to verify the distributive law for the matrix A and the vectors x and y below.

```
> A := matrix([[2,3],[5,7]]);
> x := colvector([2,3]);
> y := colvector([5,-1]);
>
```

Solution: We calculate $A\,x + A\,y$ and $A\,(x+y)$:

```
> evalm(A&*x+A&*y)=evalm(A&*(x+y));
>
```

====================

Exercise 2.2: (By hand) Let A be an m by 3 matrix with columns A_1, A_2, A_3, and let x and y be 3 by 1 vectors (see below). Use vector algebra to prove the distributive law $A\,(x+y) = A\,x + A\,y$. Hint: Use the definition of a matrix-vector product to expand the products $A\,x$, $A\,y$ and $A\,(x+y)$.

$$A = [A_1, A_2, A_3] \qquad x = \begin{bmatrix} x_1 \\ x_2 \\ x_3 \end{bmatrix} \qquad y = \begin{bmatrix} y_1 \\ y_2 \\ y_3 \end{bmatrix}$$

```
>
```

⊞ Student Workspace

⊞ Answer 2.2

```
>
```

Solution Set of a Linear System $A\,x = b$

We now know that we can express any linear system in the compact form $A\,x = b$, where A is the coefficient matrix of the system and b is the right-side vector. The corresponding homogeneous linear system is then $A\,x = 0$. In Chapters 1 and 2, we have seen that the solution sets of $A\,x = b$ and $A\,x = 0$ are closely related:

Theorem 2: If p is a particular solution of the linear system $A\,x = b$ and Span$\{ v_1, ..., v_k \}$ is the solution set of the corresponding homogeneous system $A\,x = 0$, then p + Span$\{ v_1, ..., v_k \}$ is the solution set of $A\,x = b$.

For example, in Chapter 2, we saw that the solutions of $A\,x = 0$ can be decomposed into a span of a set of vectors $\{ v_1, ..., v_k \}$ and visualized as a plane through the origin (if $k = 2$). Also we saw that the solutions of $A\,x = b$ can be visualized as a translate by a vector p of the solution set of $A\,x = 0$. Now we can use the rules of algebra to prove Theorem 2. The proof in the following example for $k = 2$ can be extended easily to the general case.

[>

====================

Example 2C: Suppose p is a solution of $A\,x = b$ and v_1 and v_2 are each solutions of the corresponding homogeneous equation $A\,x = 0$.
(a) Use the above rules of algebra to show that $p + c_1\,v_1 + c_2\,v_2$ is a solution of $A\,x = b$ for all choices of the weights c_1, c_2.
(b) If q is another solution of $A\,x = b$ and if every solution of $A\,x = 0$ is a linear combination of v_1 and v_2, $c_1\,v_1 + c_2\,v_2$, use the above rules of algebra to show that q can be written in the form $p + c_1\,v_1 + c_2\,v_2$.

[>

Solution: (a) By the rules of algebra, $A(p + c_1\,v_1 + c_2\,v_2) = A\,p + c_1\,A\,v_1 + c_2\,A\,v_2$. But we know that $A\,p = b$ and $A\,v_1 = 0$ and $A\,v_2 = 0$. Therefore $A(p + c_1\,v_1 + c_2\,v_2) = b + 0 + 0 = b$, and therefore $p + c_1\,v_1 + c_2\,v_2$ is a solution of $A\,x = b$.
(b) Since $A\,q = b$, by the rules of algebra $A\,(q - p) = A\,q - A\,p = b - b = 0$. Therefore $q - p$ is a solution of the homogeneous system, and so we can write $q - p$ in the form $q - p = c_1\,v_1 + c_2\,v_2$. Therefore $q = p + c_1\,v_1 + c_2\,v_2$.

====================

Exercise 2.3: (a) For the matrix A below, use matsolve(..) to solve $A\,x = 0$. Then decompose the solution set and write it in the form Span$\{ v_1, ..., v_k \}$.
(b) Use matrix-vector multiplication to check that $A\,p = b$, where p and b are also below. With no further computation, write down the entire solution set of $A\,x = b$.

```
> A := matrix([[4,1,2,-3,-2],[-2,-1,-1,4,1],[0,4,-3,2,2]]);
> p := colvector([1,0,-2,2,-3]);
> b := colvector([0,4,5]);
>
```

Student Workspace

Answer2.3

Span and the Equation $A\,x = b$

Recall that a vector b is in the *span* of the vectors $v_1, \ldots, v_k$ if there are scalars $c_1, \ldots, c_k$ such that $c_1\,v_1 + \ldots + c_k\,v_k = b$. This vector equation is the same as the matrix equation $A\,c = b$, where A is the matrix whose columns are $v_1, \ldots, v_k$ and c is the column vector with entries $c_1, \ldots, c_k$. Thus:

- The vector b is in the span of the columns of a matrix A if and only if the matrix equation $A\,x = b$ has a solution.

```
>
```

Exercise 2.4: Use the matsolve(..) command to solve the equations $A\,x = b$ and $A\,x = c$, where the matrix A and vectors b and c are defined below.
(a) Is the vector b in the span of the columns of A? If so, write b as a linear combination of the columns of A.
(b) Is the vector c in the span of the columns of A? If so, write c as a linear combination of the columns of A.

```
> A := matrix([[-3,0,4], [2,-1,5], [-4,5,2],[2,4,6]]);
> b := colvector([10, 27,-15,16]);
> c := colvector([10, 34,-15,16]);
>
```

Student Workspace

Answer 2.4

Linear Independence and the Equation $A\,x = 0$

Recall that a set of vectors $\{v_1, \ldots, v_k\}$ is *linearly independent* if the only solution of the equation $c_1\,v_1 + \ldots + c_k\,v_k = 0$ is the trivial solution where all the weights c_i are zero. By Theorem 1, solving the equation $c_1\,v_1 + \ldots + c_k\,v_k = 0$ is equivalent to solving the matrix equation $A\,x = 0$, where the columns of A are the vectors $v_1, \ldots, v_k$. Thus:

- The columns of a matrix A are linearly independent if and only if the matrix equation $A\,x = 0$ has only the trivial solution $x = 0$.

Exercise 2.5: Determine whether the columns of the matrix A below are linearly independent by using matsolve(..) to solve $A\,x = 0$.

```
> A := matrix([[4,1,3],[2,3,-1],[1,-1,2]]);
> zero := colvector([0,0,0]);
>
```

Student Workspace

Answer 2.5

Problems

Problem 1: Matrix-vector products and linear combinations

Solve the linear system $A\,x = b$ using matsolve(..), where A and b are defined below. Use the solutions to express b as a linear combination of the columns of A. Find at least three different linear combinations of these columns that equal b.

```
> A := matrix([[2,1,1,2,3], [2,0,2,2,3], [-3,3,0,-2,-3]]);
> b := colvector([5,4,-7]);
>
```

Student Workspace

Problem 2: Geometric explanation of consistency

(a) For the matrix A below, check that the linear system $A\,x = b$ is consistent for every right-side column vector b and that the solution has one free variable. Hint: Apply gausselim(..) to the augmented matrix below, which is formed by augmenting the coefficient matrix with a completely general right-side vector b.

(b) Give two separate geometric explanations for why this system is consistent and the solution has a free variable. Base your first explanation on the observation that each of the two equations in the linear system can be thought of as an equation of a plane. Base your second explanation on the observation that the columns of A are vectors in R^2 that are not all multiples of one another.

```
> A := matrix([[1,3,-2],[2,-4,1]]);
> aug := augment(A,colvector([b1,b2]));
>
```

Student Workspace

Problem 3: Geometric explanation of inconsistency

(a) For the matrix A below, check that the linear system $A\,x = b$ is inconsistent for some right-side column vectors b. Hint: Apply gausselim(..) to the augmented matrix below, which is formed by augmenting the coefficient matrix A with a completely general right-side vector b. Look closely at the last row of the reduced matrix.

```
> A := matrix([[1,2],[3,-4],[-2,1]]);
> aug := augment(A,colvector([b1,b2,b3]));
>
```

(b) Give an example of a right-side vector b for which $A\,x = b$ is inconsistent. Hint: look at the echelon form of the augmented matrix *aug*.

(c) Give a geometric explanation for why this system is inconsistent for some b's. Base your explanation on the fact that the columns of A are vectors in R^3.

Student Workspace

Problem 4: Structure of solution sets

Use only matrix-vector multiplication to check that p is a solution of the linear system $A\,x = b$, and u, v are solutions of the corresponding homogeneous system. (The coefficient matrix A, the right-side vector b, and the solutions p, u, v are given below.) In answering the questions below, do not use matsolve(..) or backsolve(..) or any other method of solving linear systems.

```
> A := matrix([[-4,-4,-1,-4,3],[-4,2,2,0,1],[2,4,2,2,-2]]);
> b := colvector([2,5,2]);
> p := colvector([-1,-3,5,0,-3]);
> u := colvector([1,-6,8,3,0]);
  v := colvector([2,3,-2,0,6]);
> 
```

(a) Write down three more solutions of $A\,x = 0$, and use matrix-vector multiplication to check that they are indeed solutions.

(b) Write down three corresponding solutions of $A\,x = b$, and use matrix-vector multiplication to check that they are indeed solutions.

Student Workspace

Problem 5: Linear systems and matrix equations

For the linear system below do the following:

(a) Express this system as a matrix equation, and use matsolve(..) to solve the matrix equation.

(b) Write down the vector equation that corresponds to this linear system. What vector have you thereby expressed as a linear combination of other vectors?

(c) Use matsolve(..) to solve the corresponding homogeneous system.

(d) Write down the vector equation that corresponds to this homogeneous system. What set of vectors have you thereby shown is linearly independent or linearly dependent? Which is it?

```
> eqn1 := 2*x1 - 2*x2 - 4*x3 = -2;
> eqn2 := 2*x1 - x2 - x3 = 2;
> eqn3 := -3*x1 +5*x2 +4*x3 = 3;
> eqn4 := -x1 - 5*x2 + 4*x3 = -3;
> A := genmatrix([eqn1,eqn2,eqn3,eqn4],[x1,x2,x3]);
> 
```

Student Workspace

Problem 6: Linear independence and dependence

Confirm that the set of vectors $\{ v_1, v_2, v_3, v_4, v_5 \}$ is linearly dependent. Which of the following subsets is linearly independent? $\{ v_1, v_2 \}$, $\{ v_1, v_2, v_3 \}$, $\{ v_1, v_2, v_5 \}$, $\{ v_1, v_2, v_3, v_4 \}$. Answer all these questions by solving a matrix equation of the form $A\,x = 0$. Explain, using the definitions of linear independence and dependence, why your answers are correct.

```
> v1 := colvector([-4, -2, -1, 4, 7]);
  v2 := colvector([7, 1, -2, -6, -8]);
  v3 := colvector([-3, 3, 0, 2, 4]);
  v4 := colvector([-6, -6, -1, -2, 3]);
```

```
v5 := colvector([1, 3, 4, -2, -6]);
>
```

Student Workspace

Pencil and Paper Practice

Note: Maple is not required for any of the problems in this section. Do all of these by hand.

1. Suppose M is a 5 by 5 matrix and u, v, w and z are the vectors below. Describe each of the matrix products $M\,u$, $M\,v$, $M\,w$ and $M\,z$ in terms of the columns of M.

$$u = \begin{bmatrix} 0 \\ 0 \\ 0 \\ 4 \\ 0 \end{bmatrix} \quad v = \begin{bmatrix} 2 \\ 0 \\ 0 \\ 1 \\ 0 \end{bmatrix} \quad w = \begin{bmatrix} 0 \\ 0 \\ 1 \\ 1 \\ 1 \end{bmatrix} \quad z = \begin{bmatrix} 1 \\ 1 \\ 1 \\ 1 \\ -1 \end{bmatrix}$$

2. (a) Suppose Q is a 30 by 6 matrix whose entries represent the grades of 30 students on 6 quizzes. For example the [10, 4] entry is student number 10 's grade on the fourth quiz. Create a 6 by 1 vector v so that the product $Q\,v$ gives a single column with each student's average quiz grade.
(b) A second matrix N which is 30 by 3 contains the grades for two one-hour tests in the first two columns and the final exam in the last column. If the instructor for the course counts each test 25% and the final 50%, what vector w should she use so that the product $N\,w$ produces the correct weighted average for each student.

3. For the matrix A and the vector b below, consider the matrix equation $A\,x = b$.

$$A = \begin{bmatrix} 2 & 3 \\ 3 & -2 \\ 5 & 0 \end{bmatrix} \quad b = \begin{bmatrix} 1 \\ 1 \\ 1 \end{bmatrix}$$

(a) Write out the linear system that is equivalent to $A\,x = b$.
(b) Write out the vector equation that is equivalent $A\,x = b$.

4. Suppose A is a 5 by 4 matrix, b is a vector in R^5 and $p = \begin{bmatrix} -2 \\ 1 \\ 0 \\ 4 \end{bmatrix}$ is a solution of the equation $A\,x = b$.

Based on this information, which of the following statements is

True / False / Cannot be determined from information given

(a) The vector b is in the span of the columns of A.
(b) The equation $A\,x = b$ has an infinite number of solutions.
(c) The columns of A form a linearly dependent set of vectors.
(d) The equation $A\,x = 2\,b$ has a solution.
(e) The set of five vectors consisting of the columns of A and b is a linearly independent set.

Linear Algebra Modules Project
Chapter 3, Module 2

Matrix Multiplication

Purpose of this module

The purpose of this module is to acquaint you with the definition of matrix multiplication, its properties, and its uses.

Prerequisites

Some familiarity with matrices, including the sum of two matrices, the product of a scalar times a matrix, and the product of a matrix and a vector.

Commands used in this module

```
> restart:  with(linalg): with(lamp):
>
```

Tutorial

Section 1: The Product of Two Matrices

In Module 1 we defined the product of a matrix and a vector. Now we use that definition to define the product of two matrices.

Definition: Suppose A is a matrix with m rows and n columns and B is a matrix with n rows and r columns. The ***matrix product*** $A\ B$ is defined column-by-column as follows. The first column of $A\ B$ is the matrix A times the first column of B; the second column of $A\ B$ is A times the second column of B; and so on. Thus each column of $A\ B$ is a matrix-vector product formed by multiplying the matrix A by the corresponding column of B.

```
>
```

We can express this definition in symbols: Let $B_1, \ldots, B_r$ denote the columns of the matrix B. Then

$$A\ B = A\ [B_1 \quad B_2 \quad \ldots \quad B_r] = [A\ B_1 \quad A\ B_2 \quad \ldots \quad A\ B_r]$$

====================

Example 1A: Find the product $A\ B$ for the matrix A and matrix B given below.

```
> A := matrix([[1,3,-2],[2,-1,4]]);
  B := matrix([[4,3],[1,5],[2,6]]);
>
```

Solution: We calculate the product directly from the definition. First we give the names B_1 and B_2 to the columns of B:

```
> B1 := column(B,1);
  B2 := column(B,2);
```

Next we calculate each column of the product by multiplying A by the corresponding column of B:

```
> AB1 := evalm(A&*B1);
> AB2 := evalm(A&*B2);
```

Putting these two column vectors together into a single matrix gives us the product $A\,B$:

```
> augment(AB1,AB2);
```

Alternatively, Maple's evalm(..) command does this all in one step:

```
> evalm(A&*B);
>
```

====================

Exercise 1.1:

(a) Compute the product MN of the following two matrices in your head. Just imagine the product of M times each column of N, and think of each of these products as a linear combination of the columns of M.

```
> M := matrix([[2,-1,3,-2],[1,4,-1,-3],[-3,1,2,3]]);
  N := matrix([[0,1,1],[0,1,0],[0,1,1],[0,1,0]]);
>
```

(b) How would you change the matrix N so that all of the columns of $M\,N$ are identical to the third column of M?

Student Workspace

Answer 1.1

Dot Product Method for Calculating $A\,B$

If the columns of B are not as simple as in the preceding exercise, computing the product of two matrices by hand can be a bit tedious. Using the dot product method for each column product $A\,B_i$ may be faster:

- If A is a matrix with m rows and n columns, and B is a matrix with n rows and r columns, then the (i, j) entry of $A\,B$ is the dot product of the ith row of A and the jth column of B.

```
>
```

Exercise 1.2: (By hand)

(a) For the matrices A and B below, calculate the product $A\,B$ using the dot product method.

$$A=\begin{bmatrix}4 & 3\\ 2 & 1\end{bmatrix} \qquad B=\begin{bmatrix}7 & -2\\ 1 & 5\end{bmatrix}$$

(b) Calculate the product $B\,A$. Since the matrix B is now on the left, we calculate the (i, j) entry of $B\,A$ by taking the dot product of the ith row of B and the jth column of A. Notice that $A\,B \neq B\,A$.
(c) For the matrices G and H below, find the (2,1) entry of $G\,H$.

[>

$$G=\begin{bmatrix}a & b & c\\ d & e & f\\ g & h & i\end{bmatrix} \qquad H=\begin{bmatrix}j & k\\ l & m\\ n & o\end{bmatrix}$$

Student Workspace

Answer 1.2

[>

Section 2: Explorations with Matrix Products

Products of Special Matrices

In this section we look at several special types of matrices and their products. Sometimes we will want to explore the behavior of matrices with random integer entries; for this we use the randmat(..) command, which produces a different matrix every time you use it.

==================

Example 2A (Identity Matrix): We multiply a random matrix A by the special matrix K below:

```
> A := randmat(3,3);
  K := matrix([[1,0,0],[0,1,0],[0,0,1]]);
```

What is the product $A\,K$? When we multiply A times the first column of K, we get the first column of A; similarly, A times the second column of K is the second column of A; and so on. That is, $A\,K = A$. What is the product $K\,A$?

```
> evalm(A&*K);
  evalm(K&*A);
```

[>

No matter what the entries of A, the product of A and the special matrix with 1's along the main diagonal and 0's elsewhere is always A. So this special matrix is called the identity matrix, and is usually denoted by the letter I. (Although we say "the" identity matrix, there is an n by n identity matrix for every n.) The property we have just observed is similar to the defining property of the number 1:

$$A\,I = A \text{ and } I\,A = A$$

==================

Maple Note: Maple won't let us use I as the name of a matrix; it has reserved I for its own use as the symbol for the complex number $\sqrt{-1}$. If you are tempted to use I as a name, you might use *Id* instead.

Try the following command, and change the number 3 to other positive integers.

```
> Id := identmat(3);
>
```

==================

Example 2B: Now let's interchange the first and second columns of the identity matrix and multiply this new special matrix *P* by random matrices:

```
> A := randmat(3,3);
  P := matrix([[0,1,0],[1,0,0],[0,0,1]]);
> evalm(A&*P);
>
```

Note that the product *A P* interchanges the first and second columns of *A*. *P* is called a permutation matrix. (A *permutation matrix* is a matrix obtained by rearranging (i.e., permuting) the columns of the identity matrix.)

The matrix *Q* below is the permutation matrix obtained by moving each column of the 3 by 3 identity matrix to the right one column, except for the last column, which is moved to the first column:

```
> Q := matrix([[0,1,0],[0,0,1],[1,0,0]]);
```

Can you predict what we will get when we multiply a random matrix *A* by *Q* ?

```
> A := randmat(3,3);
  evalm(A&*Q);
>
```

So the reordering of the columns from *A* to *A Q* is identical to that between *I* and *Q*.

==================

Exercise 2.1: The matrix *N* below is the result of permuting the columns of the matrix *A* in a particular way. Find a permutation matrix *R* such that $A\,R = N$.

```
> A := matrix([[2,3,7,6],[8,2,1,5]]);
> N := matrix([[7,6,2,3],[1,5,8,2]]);
>
```

Student Workspace

Answer 2.1

In the next module, we will learn about the rules of algebra that apply to matrix algebra. Most of the rules of the algebra of real numbers that you are familiar with are also valid for matrix algebra, as we will see. However, unlike multiplication of real numbers, matrix multiplication is not commutative! That is, $A\,B = B\,A$ is not always true for matrices. You saw an example of this in Exercise 1.2 at the end of Section 1. In fact, it is so rarely true that a matrix product commutes that if we multiply two random matrices *A* and *B*, we will almost always observe that $A\,B \neq B\,A$. Execute the following commands several times:

```
> A := randmat(2,2);
  B := randmat(2,2);
> AB := evalm(A&*B);
  BA := evalm(B&*A);
```

Occasionally, however, a matrix product does commute, as in $A\,I = I\,A$ (see Example 2A).

A *square* matrix is a matrix with the same number of rows as columns. A *diagonal* matrix is a square matrix whose nondiagonal entries are all zero:

```
> K := diag(2,-3,0,4);
>
```

Maple Note: Maple won't let us use the obvious name *D* for this diagonal matrix; it has reserved *D* as the name of the differentiation operator for functions.

Exercise 2.2: Let *A* and *K* (*K* diagonal) be the matrices shown below. How is the product *A K* related to *A*? Explain why this pattern must always be true. Hint: Analyze *A K* one column at a time.

```
> A := matrix([[1,2,3],[6,2,5],[-3,4,8]]);
> K := diag(-2,3,5);
> evalm(A&*K);
>
```

Student Workspace

Answer 2.2

```
>
```

Vector-Matrix Products: *y A*

In this subsection we consider the special case of a product of two matrices in which the first matrix has just one row. Such a matrix is called a "row vector." In Module 1, we saw that when we multiply a matrix *A* on the right by a column vector, we get a linear combination of the columns of *A*. Similarly, when we multiply *A* on the left by a row vector, we get a linear combination of the rows of A. Here is an example.

====================

Example 2C: We compute the product *y A*, where *y* and *A* are defined below, and compare this product with the linear combination of the rows of *A* in which the weights are the components of *y*:

```
> A := matrix([[a,b],[c,d],[e,f]]);
> y := matrix([[2,-3,4]]);
> combo := 2*row(A,1)-3*row(A,2)+4*row(A,3);
> evalm(y&*A);
  evalm(combo);
>
```

====================

The general rule (which is proved in Module 3) is:

- The vector-matrix product $y\,A$ is the linear combination of the rows of A in which the weights are the components of y.

Exercise 2.3: Suppose A is a 3 by n matrix and $y = [1, 0, 0]$ and $z = [1, 0, 1]$. Describe the products $y\,A$ and $z\,A$ in terms of the rows of A.

Student Workspace

Answer 2.3

```
>
```

Exercise 2.4: (Compare Exercise 2.2) Let A and K (K diagonal) be the matrices shown below. How is the product $K\,A$ related to A? Explain why this pattern will always be true. Hint: Analyze $K\,A$ one row at a time, and think of each of these rows as a linear combination of rows of A.

```
> A := matrix([[1,2,3],[6,2,5],[-3,4,8]]);
> K := diag(-2,3,5);
> evalm(K&*A);
>
```

Student Workspace

Answer 2.4

```
>
```

Exercise 2.5: (Compare Example 2B and Exercise 2.1) Find the 3 by 3 permutation matrix P so that the product $P\,A$ permutes the rows of A (defined below) by moving each row of A down one row, except the last row which is moved to the top.

```
> A := matrix([[1,2,3],[6,2,5],[-3,4,8]]);
>
```

Student Workspace

Answer 2.5

Exercise 2.6: Since multiplying on the left of a matrix allows us to manipulate the rows of the matrix, we ask whether it is possible to carry out an elementary row operation by matrix multiplication. Specifically, suppose we want to carry out the addrow(..) operation below, which multiplies the first row of M by -2 and adds the result to the second row of M. Find a matrix R so that this row operation is carried out in the product $R\,M$. Hint: We want the matrix R to leave the first and third rows of M unchanged and replace the second row by a certain linear combination of the rows of M.

```
> M := matrix([[1,2,5],[2,7,3],[4,3,9]]);
> M1 := addrow(M,1,2,-2);
>
```

Student Workspace

Answer 2.6

Powers of Matrices

If A is a square matrix, we define A^2 to be the product $A\,A$. Higher powers are defined similarly: A^n is the product of n copies of A.

Exercise 2.7: Let J be the n by n matrix whose entries are all ones. (In the Student Workspace, you will find a command that constructs the 4 by 4 matrix of all ones. Be sure to change the command to get matrices of other sizes.)

(a) What is J^2 ? (Your answer will depend on n.)

(b) Explain why your answer is correct for all values of n.

Student Workspace

Answer 2.7

```
>
```

Exercise 2.8: Let K be the matrix whose entries are all 1's along the diagonal just above the main diagonal and 0's elsewhere. (Such a matrix is called a Jordan matrix. In the Student Workspace, you will find a command that constructs the 4 by 4 Jordan matrix. Change the number 4 to get other sizes.)

(a) What is K^2?

(b) What is K^n for every positive integer power n?

(c) Explain why your answer to (b) is correct for all values of n.

Student Workspace

Answer 2.8

```
>
```

Problems

Problem 1: Permute the components of a vector

Let u and v be the vectors defined below. Find the permutation matrix P such that $P\,u = v$. (Thus P permutes the components of u to get the components of v.)

```
> u := colvector([1,2,3,4]);
  v := colvector([4,1,3,2]);
>
```

Student Workspace

Problem 2: Zero columns

Suppose A and B are matrices whose product $A\ B$ is defined.
(a) If the first column of B is all zeros, what is the first column of $A\ B$? Explain. Is a similar statement true for other coumns of B and $A\ B$?
(b) If the first column of $A\ B$ is all zeros but no column of B is all zeros, what can you say about the columns of A? (Hint: Think of the first column of $A\ B$ as a linear combination of the columns of A.) Is a similar statement true for other columns of $A\ B$?

Student Workspace

Problem 3: Commuting matrices

The 3 by 3 diagonal matrix K below commutes with all 3 by 3 diagonal matrices; that is, $A\ K = K\ A$ whenever A is diagonal. (In fact, diagonal matrices always commute with one another.) However, because of its special structure, K also commutes with some nondiagonal matrices. Find all the 3 by 3 matrices that commute with K. Hint: The upper-left 2 by 2 corner of K is a particularly simple matrix.

```
> diag(2,2,3);
```

Student Workspace

Problem 4: Inverse of a matrix

The command matsolve(A,b) only works when b is a column vector. But we can solve $A\ X = B$ for X when B has many columns by applying matsolve(..) to one column of B at a time.
(a) Use this idea to solve the equation $A\ X = I$, where A is the 3 by 3 matrix below and I is the 3 by 3 identity matrix. (The matrix X is called the "inverse" of the matrix A.)

```
> A := matrix([[1, 1, -1], [4, -2, -3], [-2, 2, 1]]);
> Id := identmat(3);
>
```

(b) Check your answer by computing $A\ X$. Also compute $X\ A$. (Note that the relationship between X and A is similar to the relationship between a real number and its reciprocal.)
(c) Every nonzero real number has a reciprocal, but not every nonzero matrix has an inverse. Find a nonzero 3 by 3 matrix that has no inverse. Explain why it cannot have an inverse.

Student Workspace

Problem 5: Elementary row operations

(a) For the matrix A below, find the matrices R and S that perform these row operations on A:

- $R\ A$ multiplies row 1 of A by -3 and adds it to row 2 of A;

- $S\ A$ multiplies row 1 of A by 2 and adds it to row 3 of A.

Check your answers by computing $R\ A$ and $S\ A$.

(b) Check also that $R\,S = S\,R$, and explain why the matrices R and S commute. Hint: Think of what each of these matrices does to A.

```
> A := matrix([[2,-1,3,0],[6,-3,0,2],[-4,1,-1,3]]);
> addrow(A,1,2,-3);
  addrow(A,1,3,2);
>
```

Student Workspace

Problem 6: Jordan matrices

(a) Let K be a Jordan matrix, as in Exercise 2.8. (See the jordanmat(..) command in the Student Workspace.) If A is any square matrix, describe $A\,K$ and $K\,A$. That is, what does K do to A when it is multiplied on the right of A? on the left of A? Experiment with various size matrices. Your answer must be valid for Jordan matrices of every possible size.

(b) Explain why your answer in (a) is correct by explaining what multiplying by K does to the columns and rows of A. Hint: Think of each column of $A\,K$ as a linear combination of the columns of A, and think of each row of $K\,A$ as a linear combination of the rows of A.

```
>
```

Student Workspace

Problem 7: Backward identity matrices

Let B be a "backward identity" matrix, which is a square matrix with 1's along the antidiagonal from the upper-right to lower-left corner and 0's elsewhere. (See the backidentmat(..) command in the Student Workspace.) Experiment with products $A\,B$ and $B\,A$ (where A is any square matrix) to help you answer the following questions. Your answers to the questions must be valid for backward identity matrices of every possible size.

```
>
```

(a) Describe the products $A\,B$ and $B\,A$. That is, what does B do to A when it is multiplied on the right of A? on the left of A?

(b) Explain why your answers to (a) are correct. Hint: B is a permutation matrix.

(c) What is B^2?

(d) What is B^n for every positive integer n?

Student Workspace

Problem 8: Letter L matrices

Let L be a "letter L" matrix, which is a square matrix of 0's and 1's in which the 1's form the shape of the letter L. (See the letterL(..) command in the Student Workspace.) Experiment with the product $A\,L$ (where A is any matrix) to help you answer the following questions. Your answers to the questions must be valid for letter L matrices of every possible size.

(a) Describe the product $A\,L$. That is, what does L do to A when it is multiplied on the right of A? Explain.

(b) What is L^2?

(c) What is L^n for every positive integer n?

[>

Student Workspace

Pencil and Paper Practice Questions

Note: Maple is not required for any of the problems in this section. Do all of these by hand.

1. For the matrices A and B below, compute the product $A\,B$ three ways and check that all three methods give the same answer: (a) use the dot product method; (b) compute each column of $A\,B$ as a linear combination of the columns of A; (c) compute each row of $A\,B$ as a linear combination of the rows of B.

$$A = \begin{bmatrix} 1 & 3 & -2 \\ 2 & -1 & 4 \end{bmatrix} \qquad B = \begin{bmatrix} 2 & -3 \\ 5 & 1 \\ -1 & 3 \end{bmatrix}$$

2. If K is the special matrix below and C is an arbitrary 4 by 4 matrix, describe the product $C\,K$ in terms of the columns of C and describe the product $K\,C$ in terms of the rows of C.

$$K = \begin{bmatrix} 0 & 0 & 0 & 1 \\ 1 & 0 & 0 & 0 \\ 0 & 1 & 0 & 0 \\ 0 & 0 & 0 & 1 \end{bmatrix}$$

Linear Algebra Modules Project
Chapter 3, Module 3

Rules of Matrix Algebra

Special instructions

The Tutorial for this module is designed to be done by hand without the use of Maple. Likewise, most of the Problems can and should be done without the use of Maple. However, feel free to use Maple to try examples and to check your answers. The answers to the Exercises are collected together at the end of the Tutorial.

Purpose of this module

The purpose of this module is to acquaint you with the rules of matrix algebra, give you practice using them, and help you appreciate their limitations.

Prerequisites

Some familiarity with matrices, including the sum of two matrices, the product of a scalar times a matrix, and the product of two matrices.

```
> restart: with(linalg): with(lamp):
>
```

Tutorial

Section 1: Familiar Rules of Algebra that Apply to Matrices

The rules of matrix algebra are much like the rules for the ordinary algebra of numbers but with a couple of significant differences. First let's look at the rules of algebra that are exactly like the rules of ordinary algebra.

```
>
```

Theorem 3: Matrix algebra obeys the following rules (where uppercase letters denote matrices and lowercase letters denote scalars; also, the letter O denotes the zero matrix, and the letter I denotes the identity matrix):

$$A + B = B + A \quad \text{(commutative law for addition)}$$
$$(A + B) + C = A + (B + C) \quad \text{(associative law for addition)}$$
$$(A\,B)\,C = A\,(B\,C) \quad \text{(associative law for multiplication)}$$
$$A\,(B + C) = A\,B + A\,C \quad \text{(first distributive law)}$$
$$(A + B)\,C = A\,C + B\,C \quad \text{(second distributive law)}$$

$$A + O = A \quad \text{(identity law for addition)}$$
$$A\,I = I\,A = A \quad \text{(identity law for multiplication)}$$
$$A - A = O \quad \text{(inverse law for addition)}$$

And here are rules involving scalar multiplication:

$$(c\,d)\,A = c\,(d\,A)$$
$$c(A\,B) = (c\,A)\,B = A\,(c\,B)$$
$$(c+d)\,A = c\,A + d\,A$$
$$c\,(A+B) = c\,A + c\,B$$

[>

==================

Example 1A: Perhaps the most familiar use of rules of algebra is in solving equations. For the matrices

$$A = \begin{bmatrix} -3 & 4 & 3 \\ -4 & -2 & 2 \\ 0 & 1 & -2 \end{bmatrix} \quad \text{and} \quad B = \begin{bmatrix} 4 \\ 3 \\ -2 \end{bmatrix}$$

solve the matrix equation $4\,X = 2\,B - A\,X$ for the unknown column vector X.

[>

Solution: We write out all the steps in detail, making explicit use of the rules in Theorem 3.

- (1) Rearrange terms so all terms with X are on the left. Specifically, add $A\,X$ to both sides and then use the associate, inverse, and identity laws for addition:

$$4\,X + A\,X = 2\,B$$

- (2) Write X as $I\,X$ so we can factor out X; then use the second distributive law to factor:

$$4\,I\,X + A\,X = 2\,B$$
$$(4\,I + A)\,X = 2\,B$$

- (3) The last equation above has the familiar form of a matrix equation $C\,X = K$ where

$$C = 4\,I + A = \begin{bmatrix} 1 & 4 & 3 \\ -4 & 2 & 2 \\ 0 & 1 & 2 \end{bmatrix} \quad \text{and} \quad K = 2\,B = \begin{bmatrix} 8 \\ 6 \\ -4 \end{bmatrix}$$

- (4) Solve $C X = K$ (using, for example, matsolve(C,K)) and get $X = \begin{bmatrix} -1 \\ 6 \\ -5 \end{bmatrix}$.

==================

[>

Exercise 1.1: Multiply out and simplify the following product:

$$(I + A + A^2)(I - A)$$

Student Workspace

Answer 1.1

[>

How to Prove the Rules of Algebra

We can prove the rules of algebra in Theorem 3 fairly easily by using the two rules of algebra for matrix-vector products introduced in Module 1:

$$A(x + y) = A x + A y \qquad \text{(distributive law)}$$
$$A(c\,\mathrm{x}) = \mathrm{c}\,(A\,\mathrm{x}) \qquad \text{(scalars factor out)}$$

where A is any m by n matrix, x and y are each n by 1 column vectors and c is a any real number.

The following observation shows why rules for matrix-vector products are useful in proving rules for matrix-matrix products: The jth column of the matrix-matrix product $A B$ is the matrix-vector product of A times the jth column of B. We apply this observation to three of the most significant rules in Theorem 3:

[>

First distributive law: The jth column of $A(B + C)$ is, by the preceding observation, the matrix-vector product $A(B_j + C_j)$, where B_j is the jth column of B and C_j is the jth column of C. By the above distributive law for matrix-vector products,

$$A(B_j + C_j) = A B_j + A C_j$$

The right side of the above equation is the jth column of $A B + A C$. Therefore $A(B + C) = A B + A C$.

[>

Second distributive law: This law can be proved in a similar way, but it requires a different distributive law for matrix-vector products:

$$(A + B)x = A x + B x$$

Although we will omit the proof of this distributive law, it is not difficult and can be proved in the same straightforward way as the other rules of algebra for matrix-vector products.

[>

Associative law for multiplication: This law requires yet another new rule for matrix-vector products, namely the following associative law:

$$(A\ B)\ x = A\ (B\ x)$$

Here's a proof of this rule, where x_j denotes the jth component of the vector x and B_j denotes the jth column of B and P_j denotes the jth column of the product $A\ B$:

$$(A\ B)\ x = x_1\ P_1 + \ldots + x_k\ P_k = x_1\ A\ B_1 + \ldots + x_k\ A\ B_k = A\ (x_1\ B_1 + \ldots + x_k\ B_k) = A\ (B\ x)$$

Note especially the third equal sign in the above proof; at that step we used both of the two rules of algebra for matrix-vector products stated at the beginning of this subsection.

[>

Section 2: Familiar Rules of Algebra that Do Not Apply to Matrices

Some rules that are true for the ordinary algebra of numbers are false for matrix algebra. In Module 2, for example, we observed that usually $A\ B \neq B\ A$ when A and B are matrices. This fact is usually put into words as follows: Although multiplication of numbers is commutative, matrix multiplication is not commutative. Here is an example from Module 2 of a pair of matrices that do not commute:

$$\begin{bmatrix} 4 & 3 \\ 2 & 1 \end{bmatrix} \begin{bmatrix} 7 & -2 \\ 1 & 5 \end{bmatrix} \neq \begin{bmatrix} 7 & -2 \\ 1 & 5 \end{bmatrix} \begin{bmatrix} 4 & 3 \\ 2 & 1 \end{bmatrix}$$

[>

While in general matrix multiplication is not commutative, examples of matrices that do commute are fairly easy to find. We have already seen that the identity matrix commutes with any matrix. The next two examples consider some other special matrices that commute.

==================

Example 2A: For the diagonal matrix $K = \begin{bmatrix} 2 & 0 \\ 0 & -3 \end{bmatrix}$, what are the matrices A such that $K\ A = A\ K$?

Hint: Think of how K changes A when it is multiplied on the left and right of A.

Solution: We saw in Module 2 that diagonal matrices always commute. So $K\ A = A\ K$ for every diagonal matrix A. Do any nondiagonal matrices commute with K? $K\ A$ multiplies the first row of A by 2 and the second row by -3, and $A\ K$ multiplies the first column of A by 2 and the second column by -3. So $K\ A = A\ K$ if and only if both operations produce the same matrix. This is true only when A is also a diagonal matrix.

==================

[>

Exercise 2.1: For the permutation matrix $P = \begin{bmatrix} 0 & 1 \\ 1 & 0 \end{bmatrix}$, what are the matrices A such that $P\,A = A\,P$?

Hint: Think of how P changes A when it is multiplied on the left and right of A.

Student Workspace

Answer 2.1

[>

Here is another way in which the rules of matrix algebra differ from those of ordinary algebra:

==================

Example 2B: If $A = \begin{bmatrix} 1 & 1 \\ 1 & 1 \end{bmatrix}$, find a 2 by 2 matrix B such that $A\,B = \mathrm{O}$ (where O denotes the matrix of all zeros) but no entry of B is zero.

Solution: One answer is $B = \begin{bmatrix} 1 & 1 \\ -1 & -1 \end{bmatrix}$. How can you find other answers? Hint: Solve the equation

$$\begin{bmatrix} 1 & 1 \\ 1 & 1 \end{bmatrix}\begin{bmatrix} a & b \\ c & d \end{bmatrix} = \begin{bmatrix} 0 & 0 \\ 0 & 0 \end{bmatrix}$$

by converting this equation into four linear equations in the four unknowns a, b, c, d.

==================

[>

So the familiar cancellation law of ordinary algebra is false for matrices: $A\,B = \mathrm{O}$ does <u>not</u> imply that $A = \mathrm{O}$ or $B = \mathrm{O}$.

Summary: The following are the main differences between the rules of matrix algebra and the rules of ordinary algebra:

- Matrix multiplication is not commutative.
- There is no cancellation law for matrix products.

[>

Section 3: Rules for the Transpose of a Matrix

The ***transpose*** of a matrix A is the matrix obtained from A by making its rows into its columns (and hence its columns into its rows). The Maple command for the transpose of A is transpose(A); the mathematical symbol for it is A^T. Here are some matrices and their transposes:

$$A = \begin{bmatrix} 2 & 3 & 4 \\ 6 & 8 & 9 \end{bmatrix}, \quad A^T = \begin{bmatrix} 2 & 6 \\ 3 & 8 \\ 4 & 9 \end{bmatrix}$$

$$B = \begin{bmatrix} 12 & 35 & 0 \\ 68 & 94 & -56 \\ 3 & 6 & 8 \end{bmatrix}, \quad B^T = \begin{bmatrix} 12 & 68 & 3 \\ 35 & 94 & 6 \\ 0 & -56 & 8 \end{bmatrix}$$

[>

Note that for a square matrix (same number of rows as columns), its transpose can found by reflecting the entries of the matrix across the main diagonal.

Most of the rules of algebra for the transpose are straightforward and easy to use, but the fourth rule below may surprise you:

[>

Theorem 4: The transpose operation for matrices obeys the following rules of algebra (where uppercase letters denote matrices and the lowercase letter is a scalar):

$$(A + B)^T = A^T + B^T$$
$$(c\,A)^T = c\,A^T$$
$$(A^T)^T = A$$
$$(A\,B)^T = B^T A^T$$

Note that in the fourth rule the order of the matrices A and B is switched when you take the transpose of their product. Thus, in general, $(A\,B)^T \neq A^T B^T$.

[>

Exercise 3.1: Suppose A is a 2 by 2 matrix and B is a 2 by 3 matrix. Explain how, without doing any calculating, you can tell that $(A\,B)^T$ could not possibly equal $A^T B^T$ for these particular matrices. (Hint: look at the size of each of the matrices.)

Student Workspace

Answer 3.1

[>

Exercise 3.2:

(a) What entry of A equals the $(i,\ j)$ entry of A^T? (Hint: First consider a specific entry of A, such as the (1, 2) entry.)

(b) How is the (i, j) entry of $(A\,B)^T$ calculated from the entries of A and B? (Hint: The (i, j) entry of $A\,B$ is the dot product of the ith row of A and the jth column of B.)

Student Workspace

Answer 3.2

[>

==================

Example 3A: Solve the equation $2\,(A\,X)^T = X^T\,M + B^T\,A^T$ for the column vector X, given the matrices A, B, M below.

$$A = \begin{bmatrix} 2 & 3 \\ 5 & 7 \end{bmatrix},\quad B = \begin{bmatrix} -3 \\ 2 \end{bmatrix},\quad M = \begin{bmatrix} 3 & 5 \\ 4 & 2 \end{bmatrix}$$

[>

Solution:

- (1) Take the transpose of both sides and apply all four parts of Theorem 4:

$$2\,A\,X = M^T\,X + A\,B$$

- (2) Collect all terms with X on one side:

$$2\,A\,X - M^T\,X = A\,B$$

- (3) Factor out X by using the second distributive law in Theorem 3:

$$(2\,A - M^T)\,X = A\,B$$

- (4) We now have a matrix equation in the familiar form $C\,X = K$, where

$$C = 2\,A - M^T = \begin{bmatrix} 1 & 2 \\ 5 & 12 \end{bmatrix},\quad \mathrm{K} = A\,B = \begin{bmatrix} 0 \\ -1 \end{bmatrix}$$

[>

- (5) Solve $C\,X = K$ (using, for example, matsolve(C,K)) and get $X = \begin{bmatrix} 1 \\ -\frac{1}{2} \end{bmatrix}$.

==================

====================

Example 3B: Recall the assertion made in Module 2: The product $y\,A$, where y is a row vector and A is a matrix, can be written as the linear combination of the rows of A in which the weights are the components of y. Here is an example in which we compute $y\,A$ both ways:

$$y\,A = [2 \quad -3 \quad 4] \begin{bmatrix} a & b \\ c & d \\ e & f \end{bmatrix} = [2\,a - 3\,c + 4\,e \quad 2\,b - 3\,d + 4\,f]$$

$$2\,[a \quad b] - 3\,[c \quad d] + 4\,[e \quad f] = [2\,a - 3\,c + 4\,e \quad 2\,b - 3\,d + 4\,f]$$

We can use the rules of transposes to prove the above assertion in general. First use the rule for the transpose of a product:

$$(y\,A)^T = A^T\,y^T$$

The product $A^T\,y^T$ is a matrix-vector product and is therefore the linear combination of the columns of A^T (i.e., the rows of A) in which the weights are the components of y^T (i.e., the components of y). Since $y\,A$ is the tranpose of $A^T\,y^T$, $y\,A$ is also the linear combination of the rows of A in which the weights are the components of y; but now the rows of A are written horizontally, since the transpose operation changes column vectors to row vectors.

====================

[>

Answers to Exercises

Answer 1.1:

$(I + A + A^2)\,(I - A) = (I + A + A^2)\,I - (I + A + A^2)\,A = I + A + A^2 - (A + A^2 + A^3) = I - A^3$

Answer 2.1:

$P\,A$ swaps the rows of A, and $A\,P$ swaps the columns of A. So $P\,A = A\,P$ if and only if swapping the rows of A and swapping the columns of A produce the same matrix. That is true only when A has the form $A = \begin{bmatrix} a & b \\ b & a \end{bmatrix}$.

Answer 3.1:

A^T is 2 by 2 and B^T is 3 by 2. So the product $A^T\,B^T$ is not even defined.

Answer 3.2:

(a) The (i, j) entry of A^T is the (j, i) entry of A.

(b) The (i, j) entry of $(A\,B)^T$ is the (j, i) entry of $A\,B$, which is computed as the dot product of the jth row of A and the ith column of B. But that is the same as the dot product of the ith row of B^T and the jth column of A^T. Note that, by this discussion, we have proved the rule $(A\,B)^T = B^T\,A^T$.

[>

Problems

Problem 1: Use matrix algebra to solve an equation

Express the matrix equation $A\,X + 5\,A = A\,B + 4\,X$ in the form $C\,X = K$, where C and K are expressions in A and B.

Student Workspace

Problem 2: Factor a matrix expression

Factor the matrix expression $3\,A - A\,B + 12\,C - 4\,C\,B$ into a product of two matrices $M\,N$.

Student Workspace

Problem 3: Counter-examples in matrix algebra I

These are examples that demonstrate again how different matrix algebra can be from ordinary algebra. All of the statements below are FALSE. For each statement, use legitimate rules of matrix algebra to rewrite (by hand) the given equation in a form that shows more clearly why the statement is false. Then use an example from either this module or the module on "Matrix Multiplication" to demonstrate that the statement is false.

(a) $(A - B)\,(A + B) = A^2 - B^2$

(b) $(A + B)^2 = A^2 + 2\,A\,B + B^2$

(c) If $A\,C = A\,B$ and $A \neq O$, then $C = B$.

```
>
```

Student Workspace

Problem 4: Counter-examples in matrix algebra II

These are examples that demonstrate again how different matrix algebra can be from ordinary algebra. All of the statements below are FALSE. For each part below, give an example of a matrix that FAILS to satisfy the stated property, hence demonstrating that the statement is false. (Suggestions: Try examples from either this module or the module on "Matrix Multiplication", or experiment with some simple 2 by 2 matrices whose entries are small integers, such as 0, 1, and -1.)

(a) If $A^2 = O$, then $A = O$ (O denotes the matrix of all zeros)

(b) If $A^2 = I$, then $A = I$ or $A = -I$.

(c) The entries of A^2 are all greater than or equal to 0.

Student Workspace

Problem 5: Use rules of transposes to solve a matrix equation

With the help of the rules of matrix algebra, solve the matrix equation $3\,C + 2\,X^T = X^T\,B$ for X, where B and C are given below. That is, by hand write the equation in a form that can be solved using matsolve(..). Then solve it by any method. Check your answer by substituting your solution X back into the given equation.

```
> B := matrix([[-1,3,2],[-1,3,3],[0,2,-1]]);
  C := matrix([[1,-1,4]]);
```

$$B := \begin{bmatrix} -1 & 3 & 2 \\ -1 & 3 & 3 \\ 0 & 2 & -1 \end{bmatrix}$$

$$C := \begin{bmatrix} 1 & -1 & 4 \end{bmatrix}$$

Student Workspace

Problem 6: Symmetric matrices

A matrix A is said to by *symmetric* if $A^T = A$. In other words, each row of A has the same entries as the corresponding column of A. Also note that when a symmetric matrix is reflected across its main diagonal, the matrix is unchanged. A symmetric matrix is always square.

(a) Give an example of a 3 by 3 symmetric matrix. Check that it is symmetric by applying the transpose(..) command to it. Can you find an example of a 3 by 3 symmetric matrix with 6 different entries? with 7?

(b) Use the rules of algebra for the transpose operation to prove that for every matrix A, both $A\,A^T$ and $A^T A$ are symmetric.

(c) Use the rules of algebra for the transpose operation to prove that if A and B are symmetric n by n matrices, so is $s\,A + t\,B$ for all scalars s and t.

(d) Give an example of 2 by 2 symmetric matrices A and B such that $A\,B$ is not symmetric. Hint: Use the rules of algebra to find out how A and B would have to be related if the product $A\,B$ were symmetric.

Student Workspace

>

Linear Algebra Modules Project
Chapter 3, Module 4

Markov Chains -- An Application

Purpose of this module

The purpose of this module is to introduce an important class of applications that depend on products and powers of matrices. We will return to these applications in later modules where you will be able to use your expanding knowledge of linear algebra to derive even more insights than we can at this early stage.

Prerequisites

Matrix algebra; solution of linear systems of equations.

Commands used in this module

```
> restart:  with(linalg):  with(lamp):
>
```

Tutorial

Section 1: A Typical (though simplified) Markov Chain Problem

In this section you will investigate a simple mathematical model for population migration. Our focus will be on the movement of a country's population between urban and rural areas. The table below shows current urban/rural breakdowns for a sampling of countries.

	Urban	*Rural*
Argentina	*87%*	*13%*
China	*29%*	*71%*
Ireland	*57%*	*43%*
Lampland	*45%*	*55%*
Nigeria	*16%*	*84%*
Sri Lanka	*22%*	*78%*
Sweden	*83%*	*17%*
United States	*75%*	*25%*

```
>
```

While there is clearly a wide variation among countries in the relative size of their urban populations, the population of the world as a whole is steadily moving towards the cities. In 1800, only 3% of the population of the world lived in cities. By 1900 that number had grown to 10%. As we approach the year 2000 it is projected that at least 50% of the world population will live in urban areas.

We now turn our attention to the newly formed country of Lampland. Our goal is to build a mathematical model that we can use to predict how the urban population for Lampland will change over the next 100 years. Suppose we have collected enough data to justify the following assumptions about the Lampland population:

- (1) Each year 6% of the urban population moves to rural areas and the remaining 94% stay in urban areas.
- (2) Each year 9% of the rural population moves to urban areas and the remaining 91% stay in rural areas.
- (3) At the beginning of our study, the Lampland population is 45% urban and 55% rural.

[>

This is obviously a highly simplified mathematical model of a real problem. You might ask yourself what the most serious limitations of this model are. That is, in what ways is it unrealistic, and what might we want to add to our model to overcome its most serious limitations? However, when constructing a mathematical model of a real problem, starting with a very simple model is a useful strategy. And, as we will see, even this simple model provides some useful insights.

Here are the questions we will investigate:

- (a) What will happen to the population in the long run? For example, will all of the population end up in the cities? Will the urban and rural populations eventually settle down, or will they oscillate back and forth endlessly?
- (b) How does the long-term distribution of the two populations depend on the initial 45-55 distribution? For example, if the urban population had started out at 75%, how different would the long-term distribution be?

[>

We will use u_k to denote the percent of the total population that is in urban areas after k years and r_k to denote the percent of the population that is in rural areas after k years. So, in particular, by our assumption (3), we have

$$u_0 = .45 \quad \text{and} \quad r_0 = .55$$

We can use our assumptions (1) and (2) to write equations that describe the population percentages in year $k + 1$ in terms of the population percentages in year k:

$$u_{k+1} = .94\, u_k + .09\, r_k$$
$$r_{k+1} = .06\, u_k + .91\, r_k$$

We will refer to these as our "migration equations."

[>

Derivation of the Migration Equations

Let the populations for the urban and rural areas after k years be U_k and R_k, respectively. Then by our assumptions (1) and (2), the populations for these areas for the k+1 st year will be:

$$U_{k+1} = .94\, U_k + .09\, R_k$$
$$R_{k+1} = .06\, U_k + .91\, R_k$$

To convert to population *percentages,* we divide the equations by the total population T (which we assume stays constant):

$$\frac{U_{k+1}}{T} = .94\,\frac{U_k}{T} + .09\,\frac{R_k}{T}$$
$$\frac{R_{k+1}}{T} = .06\,\frac{U_k}{T} + .91\,\frac{R_k}{T}$$

These are our migration equations, which we see from the fact that $u_k = \frac{U_k}{T}$ and $r_k = \frac{R_k}{T}$, and similarly for u_{k+1} and r_{k+1}.

[>

The migration equations are called *recursion* equations, which means that they describe how to get the next values of the variables from the previous values. We will find it easier to work with these two recursion equations if we change them into a single recursion equation using matrices and vectors. Let x_k be the vector with components u_k and r_k:

$$x_k = \begin{bmatrix} u_k \\ r_k \end{bmatrix}$$

We will call x_k the kth *state vector.*

Exercise 1.1: Write the two recursion equations above as a single recursion equation of the form

$$x_{k+1} = M\, x_k$$

for some 2 by 2 matrix M. That is, find the matrix M. Check that your matrix M is correct by computing u_1 and r_1 two ways: by computing the matrix-vector product $M\, x_0$, and by using our original two recursion equations with $k = 0$.

[>

Student Workspace

Answer 1.1

In Section 2, we will see how to use the matrix-vector form of the recursion equations, $x_{k+1} = M\, x_k$, to investigate the questions we posed above.

[>

Section 2: Analysis of the Problem

Let's restate the mathematical setup from the end of Section 1. We have a 2 by 2 matrix M and a sequence of state vectors, x_0, x_1, x_2, ..., that are related by the recursion equation

$$x_{k+1} = M\, x_k$$

where

```
> x[0] := colvector([.45,.55]);
  M := matrix([[.94,.09],[.06,.91]]);
```

Therefore, by the recursion equation, we can compute the value of x_1 from x_0, and x_2 from x_1, and so on:

```
> x[1] := evalm(M&*x[0]);
> x[2] := evalm(M&*x[1]);
> x[3] := evalm(M&*x[2]);
> 
```

Notice that we can compute x_2 another way:

$$x_2 = M\, x_1 = M\, M\, x_0 = M^2\, x_0$$

That is, x_2 can be computed directly from x_0 and M by simply multiplying x_0 by M^2.

Exercise 2.1: (a) (By hand) Produce a similar formula for computing x_3 directly from x_0 and M. Do the same for x_k.
(b) (Using Maple) Check that your new formula for computing x_3 yields the same answer as we found above for x_3. Use your new formula to compute x_9.

Student Workspace

Answer 2.1

Investigation of Question (a)

- (a) What will happen to the population in the long run? For example, will all of the population end up in the cities? Will the urban vs. rural populations eventually settle down, or will they oscillate back and forth endlessly ?

Do you see any trend yet? Formulate a hypothesis (that is, make a guess!) about the long-term trend before you go on to Exercise 2.2. Enter your hypothesis here and compare yours with another student's:

```
> 
```

Exercise 2.2: (a) Use the formula in your answer to Exercise 2 to compute x_k for some larger values of k, such as $k = 25, 50, 100$.
(b) Describe in words the long-term trend. That is, answer our question (a) in Section 1.

Student Workspace

Answer 2.2

Here is a less tedious way to see the long-term trend: a short computer program (in fact, a simple "loop") that computes the population percentages over a 150 year period in ten-year increments:

```
> for k from 10 to 150 by 10
  do evalm(M^k&*x[0]) od;
>
```

(You might want to delete the output from this command when you are done looking at it, since the volume of output will clutter up your worksheet. To do so, highlight any portion of the output, and then select "Remove Output / From Selection" in the Edit menu.)

A mathematical way in which to say that the state vectors x_k are stabilizing as k gets larger is to say that the state vectors x_k are converging to a *limit vector* x. Judging from our above computations of x_k for large k's, we might guess that the limit vector x is approximately $\begin{bmatrix} .60 \\ .40 \end{bmatrix}$. This limit vector is called a *steady-state* vector for the following reason. As the computation in the next input region shows, the limit vector x satisfies the equation $M\,x = x$. This equation says that if the urban and rural population percentages should happen to equal 60% and 40% (the components of x), these percentages will remain unchanged (i.e., "steady") from that point on.

```
> x := colvector([.6,.4]);
  M := matrix([[.94,.09],[.06,.91]]);
> evalm(M&*x);
>
```

Exercise 2.3: Explain why the limit vector x should satisfy the equation $M\,x = x$. (A plausible explanation will do.)

Student Workspace

Answer 2.3

```
>
```

So a second method for finding the steady-state vector x is to solve the equation $M\,x = x$. This equation, however, has infinitely many solutions, which makes us ask how we can pick out the one solution that we want, the steady-state vector. The answer comes from an important observation, which you may have already made. The components of all the state vectors add to 1, since they are percentages which must add to 100%. So their limit vector, x, must have this same property -- its components add to 1.

Exercise 2.4: Solve the equation $M\,x = x$. Hint: Write x as $I\,x$, and then write $M\,x = x$ as a homogeneous system of linear equations. Also use the above observation to pick out the steady-state vector from the infinitely many solutions of the homogeneous system.

```
> M := matrix([[.94,.09],[.06,.91]]);
>
```

Student Workspace

Answer 2.4

Investigation of Question (b)

- (b) How does the long-term distribution of the two populations depend on the initial 45-55 distribution? For example, if the urban population had started out at 75%, how different would the long-term distribution be?

Exercise 2.5: (a) Find the long-term behavior of x_k for several different values of x_0. For example, use

$$x_0 = \begin{bmatrix} .50 \\ .50 \end{bmatrix}, \quad x_0 = \begin{bmatrix} .20 \\ .80 \end{bmatrix}, \quad x_0 = \begin{bmatrix} .75 \\ .25 \end{bmatrix}$$

(Suggestion: For each x_0, compute x_k for some large values of k.)
(b) Describe in words your answer to question (b).

```
[ >
```

Student Workspace

Answer 2.5

One way to get some insight into the behavior of the long-term distribution of the populations is to recall that $x_k = M^k\, x_0$ and to therefore study M^k instead of x_k. Execute the following loop, which computes M^k for values of k from 10 to 150 in steps of 10:

```
> for k from 10 to 150 by 10
  do evalm(M^k) od;
> 
```

So the powers M^k also approach a limit N, and both of the columns of the limit matrix appear to equal the steady-state vector x. If this is true, then, no matter what initial vector x_0 we choose, the limit vector will always be $N\, x_0$, where $N = [x, x]$.

```
> x := colvector([.6,.4]);
  N := augment(x,x);
> 
```

Exercise 2.6: (By hand) Calculate the product $N\, x_0$, where x_0 is any initial vector whose entries are percents that add to 100%, and show that the product is alway x. (Therefore the state vectors $M^k\, x_0$ always converge to the limit vector x, which depends only on M and not on x_0.)

Student Workspace

Answer 2.6

```
[ >
```

Section 3: General Observations About Markov Chains

The type of mathematical model that we used for our migration problem can be applied to a large class of similar problems. The aspects of our model that generalize are: (1) the recursion relation $x_{k+1} = M\,x_k$ that links the matrix M with the state vectors x_k, and (2) the fact that the columns of M and the state vectors have the property that their components add to 1 and are non-negative. The second fact is important, since it suggests that the concept of probability is involved; probabilities are non-negative, and they add to 1 if they take into account all possible outcomes. So we are led to make the following definition, which extracts some key parts of our set-up of the problem.

```
[ >
```

> ***Definition***: A vector whose entries are non-negative numbers adding to 1 is called a ***probability vector***. A square matrix P whose columns are probability vectors is called a ***stochastic matrix*** (or a ***Markov matrix***). Suppose x_0 is a probability vector and P is a stochastic matrix, and suppose we then construct a sequence of ***state*** vectors x_1, x_2, ... by using the recursion relation $x_{k+1} = P\,x_k$. This sequence of state vectors, linked to one another by the matrix P, is called a ***Markov chain***.

The following observations that we made about our migration matrix M and our population vectors x_k can be readily deduced for every Markov chain:

- (1) The state vectors x_k can be computed directly from P and x_0 by the formula $x_k = P^k\,x_0$.
- (2) Every state vector x_k is a probability vector.

```
[ >
```

However, some of the conclusions we reached about the limit of the state vectors and the steady-state vector are not valid for every Markov chain, as the next example shows.

====================

Example 3A (Random walks): For the matrix W below, find

(a) the limit of the state vectors $x_k = \mathrm{W}^k\,x_0$ for each of the initial probability vectors u and v below;

(b) all the probability vectors x that satisfy the equation $\mathrm{W}\,x = x$ (i.e., the "steady-state vectors").

```
[ > W := matrix([[1,.5,0,0],[0,0,.5,0],[0,.5,0,0],[0,0,.5,1]]);
[ > u := colvector([0,1,0,0]);
[ > v := colvector([1,0,0,0]);
[ >
```

Solution: (a)

```
[ > for k from 10 to 150 by 10
    do evalm(W^k&*u) od;
[ > for k from 10 to 150 by 10
    do evalm(W^k&*v) od;
[ >
```

The limit of the state vectors x_k appears to be $\left[\frac{2}{3}, 0, 0, \frac{1}{3}\right]^T$ when the initial vector is $[0, 1, 0, 0]^T$, and $[1, 0, 0, 0]^T$ when the initial vector is $[1, 0, 0, 0]^T$. In fact, when $[1, 0, 0, 0]^T$ is the initial vector, every x_k equals the initial vector!

```
[ > matsolve(W-identmat(4),colvector([0,0,0,0]));
[ >
```

(b) For the above solution to be a probability vector, t_1 and t_2 must be non-negative and add to 1. Thus, the solutions of W $x = x$ that are probability vectors can be written $[a, 0, 0, 1 - a]^T$. Note that both limit vectors we found in part (a) have this form.

So, unlike the migration matrix *M*, the limiting state for the matrix *W* depends on the initial state, and there is more than one steady-state probability vector. In Exercise 3.1, you will see another difference in the behavior of these two matrices.

====================

```
[ >
```

Interpretation of the Matrix W

The matrix *W* in Example 3A is an example of a "random walk" matrix, which has the following interpretation. Suppose you are on one of the four blocks pictured below and you walk left or right at random, either moving to an adjacent block or standing still in each time increment. If you are on one of the middle two blocks, we assume that you will move either right or left with equal probability. If you are on one of the two end blocks, we assume that you remain standing there. The numbers in the picture indicate the probabilities governing your movements.

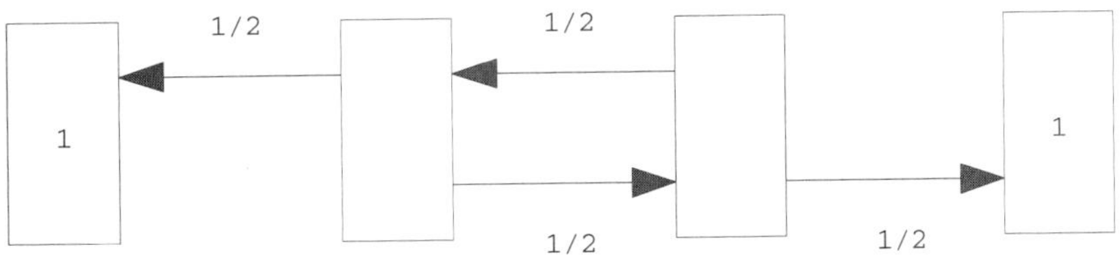


```
[ >
```

Let $x_k = [a_k, b_k, c_k, d_k]^T$ be the probability vector that represents your position after *k* time increments. That is, a_k is the probability that you are on the first (i.e., leftmost) block after *k* time increments; b_k is the probability that you are on the second block after *k* time increments, etc. The vectors x_k are then state vectors in the Markov chain linked by the stochastic matrix *W* above.

The initial state $x_0 = [0, 1, 0, 0]^T$ means that your initial position is on the second block, and the corresponding limit state, $\left[\frac{2}{3}, 0, 0, \frac{1}{3}\right]^T$, means that two-thirds of the time you will eventually land on the leftmost block and one-third of the time you will eventually land on the rightmost block.

[>

Exercise 3.1: (a) Find the limit of the state vectors x_k in Example 3A, if the initial vector is the vector x_0 below.

(b) Find the limit of the matrices W^k. Are all the columns of the limit matrix equal?

Student Workspace

Answer 3.1

In Problem 4, where we change only slightly our assumptions about the random walk, we will find yet another behavior for a Markov chain.

[>

Section 4: Regular Stochastic Matrices

In this section we look at conditions which guarantee that a stochastic matrix behaves like the migration matrix M rather than the random walk matrix W. That is, these conditions will guarantee that the state vectors x_k approach a limit that is independent of the initial state x_0. For this, we introduce the concept of a "regular" stochastic matrix.

A stochastic matrix P is said to be *regular* if for some power k all the entries of P^k are positive (i.e, none of the entries are zero). Our migration matrix M is therefore regular. The following fact can be proven (using methods beyond this course) for all regular stochastic matrices P:

- (3) The powers P^k converge to a limit matrix Q, and all the columns of Q are equal to the same probability vector x.

[>

From (3) we can readily deduce the following properties, which we observed in our analysis of the migration problem:

- (4) For every initial probability vector x_0, the state vectors $x_k = P^k\, x_0$ approach a limit vector, and the limit vector is the vector x that is in every column of Q and therefore does not depend on x_0.
- (5) Furthermore, the limit vector x is a solution of the equation $P\,x = x$ and is the only probability vector in the solution set of this homogeneous system.

Exercise 4.1: Confirm that the stochastic matrix P below is regular and that P^k does converge to a matrix whose columns are all the same probability vector.

```
[ > P := matrix([[0,.6,0],[1,0,.5],[0,.4,.5]]);
[ >
```

Student Workspace

Answer 4.1

Problems

Problem 1: What do the entries of M^2 mean?

(a) For our migration matrix M, compute M^2 and (by hand) use the entries of M^2 to express u_2 and r_2 in terms of u_0 and r_0: $u_2 = a\,u_0 + b\,r_0$, $r_2 = c\,u_0 + d\,r_0$. That is, find the constant coefficients a, b, c, d.

(b) By examining your answer to (a), describe the meaning of each of the four entries of M^2. (These meanings should be similar to the meanings of the four entries of M. For example, the (1,2) entry of M is the percentage of the rural population that moves into urban areas in one year.)

Student Workspace

Problem 2: Markov chain of healthy/sick workers

The personnel department of a large corporation is interested in constructing a mathematical model that will help them predict work force reduction due to worker illness. After researching employee attendence records over a five year period, they reached the following conclusions: Of the workers who are healthy and show up for work on a given day, 96% will be healthy and come to work the following (work) day. On the other hand, of the workers who call in sick on a given day, 60% will call in sick the next (work) day. (We'll assume that each worker is either healthy or sick on any given day and that healthy workers go to work.)

```
[ >
```

(a) Find the stochastic matrix that represents this situation. (Let the percent of healthy workers each day be the first component of the state vectors and the percent of sick workers be the second component.)
(b) Suppose that on one Monday in the middle of winter 15% of the workers call in sick. According to the assumptions of the model, what percent will be sick on Tuesday? on Wednesday? on Thursday? What percent will be sick ten weeks later?
(c) Find the steady-state vector by both of the methods we used in the Tutorial.

Student Workspace

Problem 3: Markov chain of planes at three hubs

The United Package Service has a fleet of 240 planes that shuttle between three national distribution hubs: Seattle, Chicago and Atlanta . Each day, most of the planes fly round trips returning to their starting location; however, some remain at the other two airports. Here are the percentages that move from each location:

[>

- Each day 2% of the planes from Seattle remain at Chicago and 8% remain at Atlanta; the rest return to Seattle;
- Each day 3% of the planes from Chicago remain at Seattle and 6% remain at Atlanta; the rest return to Chicago;
- Each day 9% of the planes from Atlanta remain at Seattle and 4% remain at Chicago; the rest return to Atlanta.

(a) Find the stochastic matrix that represents this situation. (Let the three components of the state vectors represent the percent of the total number of planes that are in Seattle, Chicago, and Atlanta, respectively.)

(b) Assuming that the distribution of planes has reached a steady state, how many planes are at each location?

Student Workspace

Problem 4: Random walks -- a variation

In this problem we change slightly our assumptions about the random walk in Example 3A. Let's assume that, when we are on the leftmost block or the rightmost block, we always walk to the adjacent block next. Here's a picture of the probabilities:

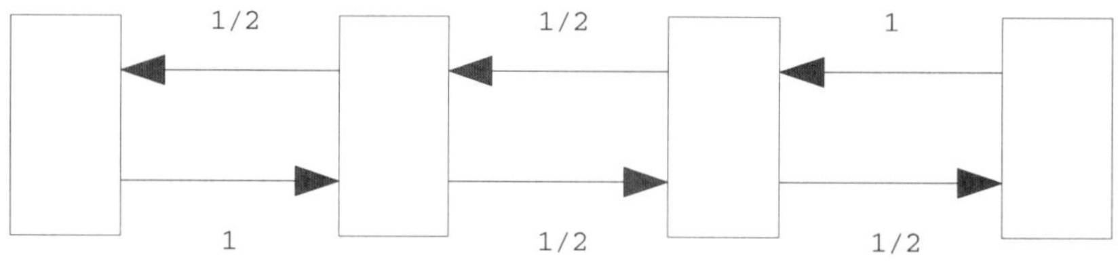

The corresponding Markov matrix is:

```
[ > W := matrix([[0,.5,0,0],[1,0,.5,0],[0,.5,0,1],[0,0,.5,0]]);
[ >
```

(a) Find all the probability vectors that satisfy the equation W $x = x$. (There is only one such vector, even though W is not regular.)

(b) Check that the state vectors x_k do not have a limit as follows. Let $x_0 = [0, 1, 0, 0]^T$, and compute W^k x_0 for several large <u>consecutive</u> values of k.

(c) Despite the fact that the x_k's have no limit, their values for large k are related to the value of the steady-state vector you found in (b). What is that relationship? Does it depend on the choice of x_0?

Student Workspace

Problem 5: Gambler's ruin

The stochastic matrix P below represents a situation that is sometimes called "Gambler's Ruin." Imagine a game of chance in which the bettor has a 60% chance of winning (and hence a 40% chance of losing). Suppose the bet for each game is $1 and that the bettor will quit if he goes broke (i.e. he reaches $0) or if he reaches his goal (which we assume is $6). The seven components of the state vector represent the following seven probabilities: the probability that the bettor has, respectively, $0, $1, $2, $3, $4, $5, $6.

[>

(a) Suppose the bettor starts with $2. (So $x_0 = [0, 0, 1, 0, 0, 0, 0]$). What is the probability that after three games the bettor has $0? $1? $2? $3? $4? $5? $6? (Do a matrix computation to get your answers.)
(b) Suppose the bettor starts with $2. What is the probability that he will eventually go broke and the probability that he will eventually reach his goal? Answer the same question for a bettor starting with $3.
(c) P is not a regular stochastic matrix. Use your answer for (b) to explain how we know this.
(d) Find all the solutions of $P\,x = x$ that are also probability vectors. Show that your answers to (b) are included among these solutions.
(e) Find the limit matrix of P^k. (That is, approximate the limit by calculating P^k for some large powers k.) Which entries of this matrix give you the answers you found in (b)? Why do they give you those answers? What other probabilities can you read off of this limit matrix?
(f) Suppose a more skillful player has a 70% chance of winning. Modify the matrix P to create the stochastic matrix representing this player's results. Find the probability that this player will eventually go broke and the probability that she will eventually reach her goal (assuming a starting bet of $2).

```
> P := matrix([[1,.4,0,0,0,0,0],[0,0,.4,0,0,0,0],
    [0,.6,0,.4,0,0,0],[0,0,.6,0,.4,0,0],[0,0,0,.6,0,.4,0],
    [0,0,0,0,.6,0,0],[0,0,0,0,0,.6,1]]);
```

[>

Student Workspace

Problem 6: Products of stochastic matrices

[>

(a) Let P be a 2 by 2 stochastic matrix and y be a probability vector with 2 components. By a direct hand computation, show that $P\,y$ is a probability vector.
(b) Let P and Q be 2 by 2 stochastic matrices. Use (a) to give a quick proof that $P\,Q$ is a stochastic matrix.

Student Workspace

[>

Pencil and Paper Practice Questions

Note: Maple is not required for any of the problems in this section. Do all of these by hand.

1. Find the unique steady-state probability vector for the regular stochastic matrix $P = \begin{bmatrix} .1 & .6 \\ .9 & .4 \end{bmatrix}$.

2. Which of the following statements is true for all stochastic matrices P, which is true for some (but not all) stochastic matrices P, and which is true for no stochastic matrices P?
(a) If y is a probability vector, so is $P\,y$.
(b) $P\,y = y$ has a unique solution.
(c) The powers P^k converge to a limit matrix.

3. Give an example of a 2 by 2 stochastic matrix that is not regular. How do you know it is not regular?

Linear Algebra Modules Project
Chapter 3, Module 5

Inverse of a Matrix - An Introduction

Purpose of this module

The purpose of this module is to introduce the concept of the inverse of a matrix and to explore its basic properties. In Section 2, we make connections between the concept of invertibility and earlier ideas from the course, such as linear independence and span of vectors.

Prerequisites

Matrix algebra, solutions of linear systems, and linear independence and span of vectors.

Commands used in this module

```
> restart:  with(linalg):  with(lamp):
>
```

Tutorial

Section 1. The Inverse of a Matrix

Throughout this module we will only consider square matrices, as these are the only matrices that have an inverse.

```
>
```

Definition: The ***inverse*** of an n by n matrix A is an n by n matrix C such that

$$A\,C = I \text{ and } C\,A = I$$

where I is the n by n identity matrix. The matrix A is said to be ***invertible*** if it has an inverse. The notation for the inverse of A is $A^{(-1)}$.

Warning: We never write $A^{(-1)}$ as $\frac{1}{A}$, since $\frac{B}{A}$ is ambiguous. (Is it $B\frac{1}{A}$ or $\frac{1}{A}B$?)

==================

Example 1A: The matrix A below has an inverse (i.e., is invertible), and we can find it by using either of the commands evalm(A^(-1)) or inverse(A).

```
> A := matrix([[-3,2],[5,4]]);
> C := evalm(A^(-1));
  C := inverse(A);
```

Let's check that C is the inverse of A by having Maple calculate the matrix products AC and CA:

```
> evalm(A&*C);
> evalm(C&*A);
>
```

==================

Here are two basic properties of the inverse:

- If C is the inverse of A, then A is the inverse of C. (In other words, the inverse of the inverse of A is A.)
- If B and C are both inverses of A, then $B = C$. (In other words, a matrix can have only one inverse.)

It is not hard to see why these two facts are true. If C is the inverse of A, then $A\,C = I$ and $C\,A = I$ by the above definition. But these two equations also say that A is the inverse of C, by the same definition. To check this, we ask Maple to compute the inverse of C. Do we get A?

```
> evalm(C^(-1));
  evalm(A);
>
```

Here is why the second fact is true. We write the product $B\,A\,C$ two ways by using the associative law of multiplication:

$$B\,A\,C = B\,(A\,C) = B\,I = B$$
$$B\,A\,C = (B\,A)\,C = I\,C = C$$

Therefore $B = C$.

```
>
```

Finding the inverse of a 2 by 2 matrix is something you can do quite easily by hand. Here's how:

The inverse of the matrix $A = \begin{bmatrix} a & b \\ c & d \end{bmatrix}$ is the matrix

$$\begin{bmatrix} \dfrac{d}{a\,d - b\,c} & -\dfrac{b}{a\,d - b\,c} \\ -\dfrac{c}{a\,d - b\,c} & \dfrac{a}{a\,d - b\,c} \end{bmatrix}$$

Maple confirms this fact:

```
> A := matrix([[a,b],[c,d]]):
  evalm(A^(-1));
>
```

Of course this formula is of no use if $a\,d - b\,c = 0$. In fact, a 2 by 2 matrix is invertible if and only if $a\,d - b\,c \neq 0$. You might recognize the number $a\,d - b\,c$ as the determinant of the matrix A. So a 2 by 2 matrix A is invertible if and only if $\det(A) \neq 0$.

Exercise 1.1: (By hand) (a) Apply the above formula to calculate the inverse N of the matrix M defined below. Then check your work by calculating the products $M\,N$ and $N\,M$.

(b) Check that the matrix S below does not have an inverse. (Apply the determinant test above, and also see what the command evalm(S^(-1)) does.)

```
> M := matrix([[6,3],[2,8]]);
> S := matrix([[2,6],[1,3]]);
>
```

Student Workspace

Answer 1.1

In Answer 1.1, note that Maple referred to the matrix S above as "singular". A matrix that does not have an inverse is called a *singular* matrix. Thus, "singular" is a synonym for "noninvertible". Likewise, "nonsingular" is a synonym for "invertible." Here it is one more time:

- If a matrix A does not have an inverse, we say that A is *noninvertible* or *singular*.

- If a matrix A does have an inverse, we say that A is *invertible* or *nonsingular*.

==================

Example 1B: If A and B are n by n invertible matrices, then it turns out that their product $A\,B$ is also invertible. Let's look at an example. The matrices A and B below each have an inverse. Notice that the product matrix $C = A\,B$ also has an inverse.

```
> A := matrix([[1,2,3],[2,1,4],[2,3,0]]);
  B := matrix([[3,2,1],[5,-3,0],[2,3,3]]);

> evalm(A^(-1));
  evalm(B^(-1));

> evalm((A&*B)^(-1));
>
```

==================

A natural question to ask at this point is whether $(A\,B)^{(-1)} = A^{(-1)}\,B^{(-1)}$?

```
> evalm(A^(-1)&*B^(-1));
>
```

No, $(A\,B)^{(-1)} \neq A^{(-1)}\,B^{(-1)}$. However, by another clever use of the associative law of multiplication, we can find a relationship among these three inverses (see Exercise 1.2).

Exercise 1.2: (By hand) Simplify the product below. From the result, state the relationship among the three inverses.

$$(A\,B)\,(B^{(-1)}\,A^{(-1)})$$

Student Workspace

Answer 1.2

Exercise 1.3: (By hand) Assume that R, S and T are three invertible matrices. Find a formula for the inverse of the product $R\,S\,T$. (Your formula should be in terms of $R^{(-1)}$, $S^{(-1)}$, and $T^{(-1)}$.) Then test your conclusion on some actual matrices.

Student Workspace

Answer 1.3

We summarize the main facts we have established for invertible matrices:

Theorem 5: Suppose A and B are n by n invertible matrices. Then:
(a) A has only one inverse matrix.
(b) $A^{(-1)}$ is invertible and $(A^{(-1)})^{(-1)} = A$.
(c) $A\,B$ is invertible and $(A\,B)^{(-1)} = B^{(-1)}\,A^{(-1)}$.

```
>
```

===================

Example 1C (Matrix equations revisited): The inverse of a matrix can be used to solve certain matrix equations, as follows. Let's assume we have a matrix equation $A\,x = b$, and suppose the matrix A happens to be a square invertible matrix. Then we can solve this equation by the following steps:

Multiply the equation $A\,x = b$ on the left by $A^{(-1)}$:

$$A^{(-1)}\,A\,x = A^{(-1)}\,b$$

Simplify the left side using the properties $A^{(-1)}\,A = I$ and $I\,x = x$:

$$x = A^{(-1)}\,b$$

Notice that when the coefficient matrix A is square and invertible, the solution set of $A\,x = b$ is a single vector, namely $A^{(-1)}\,b$. Here is an example:

```
> A := matrix([[2,3,5],[3,2,7],[3,2,4]]);
> b := colvector([3,4,1]);
> evalm(A^(-1)&*b);
```

And here, for sake of comparison, is the solution we get using matsolve(..).

```
> matsolve(A,b);
>
```

===================

Caution: The method illustrated above has two significant limitations. First, it only works if A is invertible. Second, as elegant at this method seems to be, in practice it is not an efficient way to solve a linear system, since calculating an inverse requires many more additions and multiplications than Gaussian elimination.

Exercise 1.4: Solve the linear system $A\,x = b$, where A and b are defined below. Use the method of Example 1C.

```
> A := matrix([[3,2,4],[1,3,5],[8,6,7]]);
> b := colvector([1,4,8]);
>
```

Student Workspace

Answer 1.4

Section 2. The Invertibility Theorem

This section is devoted entirely to understanding the Invertibility Theorem (abbreviated IT), which provides important connections between the concept of the inverse of a matrix, linear independence and span of vectors, and existence of solutions of linear systems:

Theorem 6 (Invertibility Theorem): Suppose A is a n by n matrix. Then the statement that A is invertible is equivalent to each of the following statements:
(a) The only solution to the homogeneous linear system $A\,x = 0$ is the trivial solution $x = 0$.
(b) The columns of A are linearly independent.
(c) The linear system $A\,x = b$ is consistent for every vector b in R^n.
(d) The columns of A span R^n.

This theorem gives us a lot of information. In part, the IT says that an invertible matrix A has all four of the properties (a), (b), (c), (d) above; furthermore, it says that if A has any one of these four properties, then A is invertible and thus A has the other three properties as well.

```
>
```

Proof that Invertibility Implies Properties (a), (b), (c), (d)

We prove here that if A is invertible, then A has all four of the properties (a), (b), (c), (d). In the next module, we complete the proof by showing that if A has any one of these four properties, then A is invertible.

(a) Since A is invertible, $A\,x = 0$ has the unique solution $x = A^{(-1)}\,0 = 0$ (see Example 1C).

(b) To prove that the columns $A_1, A_2, ..., A_n$ of A are linearly independent, we must solve the equation

$$x_1\,A_1 + x_2\,A_2 + ... + x_n\,A_n = 0$$

for the coefficients $x_1, x_2, ..., x_n$. We can rewrite the above linear combination as $A\,x = 0$, which we proved in (a) has only the trivial solution $x = 0$. Thus, the coefficients $x_1, x_2, ..., x_n$ must all be zero, which proves that the columns of A are linearly independent.

(c) Since A is invertible, the linear system $A\,x = b$ has the solution $x = A^{(-1)}\,b$ for every vector b

in R^n (see Example 1C).

(d) If b is any vector in R^n, we must show that b is a linear combination of the columns of A. We showed in (c) that $A\,x = b$ has a solution x. We then write the equation $A\,x = b$ in the form

$$x_1 A_1 + x_2 A_2 + ... + x_n A_n = b$$

which shows that b is a linear combination of the columns of A.

```
>
```

Although we are postponing the rest of the proof of Theorem 6 to the next module, in the remainder of this module we will use the IT freely in the examples, exercises, and problems.

=====================

Example 2A: Use property (a) of Theorem 6 to determine whether the matrix A below is invertible or not.

```
> A := matrix([[4,3,6],[5,8,7],[4,5,3]]);
>
```

Solution: We use matsolve(A,0) to solve the equation $A\,x = 0$, and we double check by using $A^{(-1)}$:

```
> zero := colvector([0,0,0]):
> matsolve(A,zero);
> evalm(A^(-1)&*zero);
>
```

Since $A\,x = 0$ has only the trivial solution $x = 0$, the matrix A is invertible by the IT.

=====================

Exercise 2.1: Use the result of Example 2A to answer the following question for the matrix A in Example 2A: Does $A\,x = b$ have a solution for every vector b in R^3?

Student Workspace

Answer 2.1

Exercise 2.2: Use property (b) of Theorem 6 to determine whether the matrix B below is invertible or not. Try to determine the linear independence or dependence of the columns simply by inspection.

```
> B := matrix([[1,-1,0],[-2,3,-1],[0,2,-2]]);
>
```

Student Workspace

Answer 2.2

```
>
```

Problems

Problem 1: Solve a matrix equation

Let B and C be the invertible matrices shown below. Find the matrix A that satisfies the equation

$$B\,A\,C - C = B^4 + 5\,B\,C$$

Start by solving this equation by hand. Then use your result to find A with the help of Maple. Also, test your answer by substituting A into the above equation.

```
> B := matrix([[3,2,5],[4,6,3],[2,4,3]]);
> C := matrix([[8,3,2],[4,3,5],[0,8,0]]);
>
```

Student Workspace

Problem 2: Cancellation is not always valid

(a) The following statement is FALSE. Give a counter-example using 2 by 2 matrices.
"If A and B are n by n matrices and A is not the zero matrix and $A\,B = A$, then B must be the identity matrix."
(b) Modify the above statement so that it is true. That is, modify one of the hypotheses so the conclusion is true.

Student Workspace

Problem 3: Polynomial equations and inverses

Let A be the 3 by 3 matrix defined below.

```
> A := matrix([[-1,4,-4],[1,-3,1],[1,-2,0]]);
>
```

Execute the next input region to check that A satisfies the polynomial equation $x^3 + 4\,x^2 + 5\,x + 2 = 0$. That is, $A^3 + 4\,A^2 + 5\,A + 2\,I = \mathrm{O}$. (Note that the number x is replaced by the matrix A, the number 2 by the matrix $2\,I$, and the number 0 by the zero matrix.)

```
> I3 := identmat(3):
  evalm(A^3+4*A^2+5*A+2*I3);
>
```

(a) (By hand) Rewrite the matrix equation $A^3 + 4\,A^2 + 5\,A + 2\,I = \mathrm{O}$ to find a formula for the inverse of A as a polynomial in A.
(b) Use Maple to calculate the inverse of A from your formula in (a). You may also use Maple to check your answer.

Student Workspace

Problem 4: Simplify an expression involving inverses

(By hand) Simplify the following expression, where A, B, C are n by n invertible matrices. (It simplifies drastically.)

$$(A\,B^{(-1)})^{(-1)}\;(C\,A^{(-1)})^{(-1)}\;(B\,C^{(-1)})^{(-1)}$$

Student Workspace

Problem 5: When can we solve $X\,A = B$?

Suppose A and B are n by n matrices. Sometimes we can solve the equation $X\,A = B$ for X (also n by n) and sometimes not.

(a) For the matrices A and B below, solve $X\,A = B$ for X, and describe your method of solution.

(b) Under what conditions does your method of solution in (a) work?

```
> A := matrix([[4, -4, 2], [-2, -4, 4], [0, 0, 2]]);
> B := matrix([[4, 1, 4], [1, 2, 4], [1, -4, 1]]);
>
```

Student Workspace

Problem 6: Inverse of a triangular matrix

A square matrix is "upper triangular" if all its entries below the main diagonal are zero. The matrix T below is an example.

(a) Explain why the product $T_1\,T_2$ of any two n by n upper triangular matrices is also upper triangular. (For example, what entries of T_1 and T_2 are used to compute the (2, 1) entry of $T_1\,T_2$?)

(b) By similar reasoning, the inverse of any upper triangular matrix is also upper triangular. (See T and $T^{(-1)}$ below for an example.) How are the diagonal entries of the inverse related to the diagonal entries of the matrix? Explain.

(c) There is an easy method for telling whether an upper triangular matrix is invertible without doing any computation. What is this method?

```
> T := matrix([[2,-1,3],[0,-3,5],[0,0,1/2]]);
> evalm(T^(-1));
>
```

Student Workspace

Problem 7: Inverse of A^T

If a matrix A has an inverse, then so will its transpose. In fact, these inverses are closely related. You can discover this relationship quickly by looking at a couple of examples below.

(a) Once you have found the relationship, state it in words.

(b) Prove that the relationship you found in (a) is true. Hint: Apply the transpose operation to the identity $A\,A^{(-1)} = I$, and simplify by using the rule for the transpose of a product.

```
> A := randmat(3,3);
> B := transpose(A);
> evalm(A^(-1));
  evalm(B^(-1));
>
```

Student Workspace

Problem 8: Inverse of a permutation matrix

A *permutation* matrix is a matrix obtained by rearranging the columns of an identity matrix.
(a) For the 4 by 4 permutation matrix P below, compute its transpose and inverse. How are they related?
(b) (Challenge) Explain why the relationship you described in (a) holds for all permutation matrices. Hint: Think of how P is created by permuting columns of the identity matrix, and then think of how P^T is created from the identity matrix.

```
> P := matrix([[0,0,1,0],[0,0,0,1],[0,1,0,0],[1,0,0,0]]);
>
```

Student Workspace

Problem 9: Invertible or noninvertible at a glance

For each of the matrices below, determine whether or not the matrix is invertible by inspection (i.e., doing no calculation). For each matrix, base your determination on one of the four properties of the Invertibility Theorem, and explain how you know that the matrix does or does not have that property.

```
>
```

(a) $A = \begin{bmatrix} 1 & a & b \\ 0 & 4 & c \\ 0 & 0 & 6 \end{bmatrix}$ a, b, c real numbers (Do the values of a, b, c matter?)

(b) $A = \begin{bmatrix} 1 & 2 & 3 \\ 0 & k & 4 \\ 0 & 0 & 9 \end{bmatrix}$ k a real number (Does the value of k matter?)

(c) $A = \begin{bmatrix} 1 & 2 & 7 \\ 2 & 4 & 9 \\ 3 & 6 & 11 \end{bmatrix}$

(d) A is a square matrix, and there are two vectors $u \neq v$ such that $A\,u = A\,v$.

Student Workspace

Problem 10: Invertibility and linear independence

For each of the following matrices, determine whether the matrix is invertible or not by determining whether its columns satisfy a linear dependence relation. If the columns do satisfy a linear dependence relation, find one. If you can make your determination by inspection (i.e., without any calculation) please do so.

```
> A := matrix([[1,3],[4,-1]]);
> B := matrix([[1,3],[-2,-6]]);
> C := matrix([[0, 3, -4], [-5, -2, -1], [-9, 5, 8]]);
```

```
[ > E := matrix([[-2,0,-2,-1],[2,1,1,2],[6,4,2,5],[-3,-5,2,0]]);
[ >
```

Student Workspace

Pencil and Paper Practice Questions

Note: Maple is not required for any of the problems in this section. Do all of these by hand.

1. Compute the products $A\,B$ and $B\,A$, where

$$A = \begin{bmatrix} 2 & -3 \\ -5 & 7 \end{bmatrix} \quad \text{and} \quad B = \begin{bmatrix} -7 & -3 \\ -5 & -2 \end{bmatrix}$$

From the definition of the inverse of a matrix, what can you conclude about A and B?

2. Suppose B is a square matrix such that $B^2 = 0$.
(a) Multiply out $(I - B)\,(I + B)$ and simplify.
(b) Explain why we know that $I - B$ is invertible. Write a formula for the inverse of B as a polynomial in B.

3. For each of the following matrices, determine by inspection whether the columns of the matrix are linearly independent or linearly dependent. Then use the Invertibility Theorem to determine whether the matrix is invertible or not.

$$A = \begin{bmatrix} 3 & 5 \\ 6 & -2 \end{bmatrix}$$

$$B = \begin{bmatrix} 1 & 3 & 1 \\ -2 & 0 & 1 \\ 0 & 4 & 2 \end{bmatrix}$$

$$C = \begin{bmatrix} 0 & 0 & 1 \\ 1 & 0 & 0 \\ 0 & 1 & 0 \end{bmatrix}$$

4. Suppose v_1, v_2, v_3, v_4, v_5 are vectors in R^5. How could you use a command of the form matsolve(A,0) to determine whether $\{ v_1, v_2, v_3, v_4, v_5 \}$ spans R^5? Use the Invertibility Theorem to justify your method.

Linear Algebra Modules Project
Chapter 3, Module 6

Inverse of a Matrix and Elementary Matrices

Purpose of this module

The purpose of this module is to introduce elementary matrices and use them to further our understanding of inverses. In Section 2 we take the information derived from our study of elementary matrices and use it to complete our discussion of the Invertibility Theorem (see Module 5).

Prerequisites

Matrix algebra and matrix inverses, solutions of linear systems, and linear independence and span of vectors.

Commands used in this module

```
> restart:  with(linalg):  with(lamp):
>
```

Tutorial

Section 1. Elementary Matrices

Definition: An ***elementary matrix*** is a matrix that is created by performing a single elementary row operation on an identity matrix.

====================

Example 1A: Below are three 2 by 2 elementary matrices and the elementary row operations that create them:

```
> I2 := identmat(2);
```

Row Operation: Multiply row 1 by 1/2.

```
> mulrow(I2,1,1/2);
```

Row Operation: Add -4 times row 1 to row 2.

```
> addrow(I2,1,2,-4);
```

Row Operation: Swap rows 1 and 2.

```
> swaprow(I2,1,2);
>
```

====================

Exercise 1.1: For each of the three elementary matrices *A*, *B*, *C* below, use an elementary row operation to construct the matrix from an identity matrix:

```
> A := matrix([[1,0,0],[0,1,0],[0,5,1]]);
> B := matrix([[0,0,1,0],[0,1,0,0],[1,0,0,0],[0,0,0,1]]);
> C := matrix([[1,0,0],[0,-6,0],[0, 0, 1]]);
> I3 := identmat(3):
> I4 := identmat(4):
>
```

Student Workspace

Answer 1.1

Inverses of Elementary Matrices

====================

Example 1B: We now calculate the inverses of the above elementary matrices *A*, *B* and *C*. Notice that in each case the inverse matrix is also an elementary matrix. In fact, the inverse corresponds to the "inverse" of the original row operation. By the inverse of a row operation we mean the row operation of the same type which undoes the effect of the original row operation, returning the elementary matrix to the identity matrix.

```
> evalm(A);
  evalm(A^(-1));
```

A adds 5 times row 2 to row 3, and $A^{(-1)}$ adds -5 times row 2 to row 3.

```
> evalm(B);
  evalm(B^(-1));
```

B swaps row 1 and row 3, and $B^{(-1)}$ swaps row 1 and row 3. This is an example of a matrix that is its own inverse!

```
> evalm(C);
  evalm(C^(-1));
>
```

C multiplies row 2 by -6, and $C^{(-1)}$ multiplies row 2 by -1/6.

====================

Exercise 1.2: (By hand) For each of the elementary matrices *P* and *Q* defined below, describe in words the elementary row operation that created it and describe the inverse of this elementary row operation. Use the latter to find the inverse of the elementary matrix. You can check your work by having Maple calculate the inverse.

```
> P := matrix([[1,0,-1/4,0],[0,1,0,0],[0,0,1,0],[0,0,0,1]]);
> Q := matrix([[1,0,0,0,0],[0,0,0,1,0],[0,0,1,0,0],
  [0,1,0,0,0],[0,0,0,0,1]]);
>
```

Student Workspace

Answer 1.2

Elementary Matrices Perform Row Operations

====================

Example 1C: Let A be a general 2 by 2 matrix, and let E be the elementary matrix formed by adding 5 times row 1 to row 2:

```
> A := matrix([[a,b],[c,d]]);
> I2 := identmat(2):
  E := addrow(I2,1,2,5);
```

Now we multiply A on the left by E.

```
> evalm(E&*A);
```

This seems to have performed the same row operation on A as the row operation that created E. Execute the next line to confirm this.

```
> addrow(A,1,2,5);
>
```

====================

Exercise 1.3: Below is a general 3 by 2 matrix B and two elementary matrices F and G.

(a) Without doing any calculation, predict what the product $F\,B$ will equal.

(b) Similarly, predict what the product $G\,F\,B$ will equal.

```
> B := matrix([[a,b],[c,d],[e,f]]);
  F := matrix([[1,0,0],[0,1,0],[0,0,-2]]);
  G := matrix([[0,0,1],[0,1,0],[1,0,0]]);
>
```

Student Workspace

Answer 1.3

The following statement describes this key property of elementary matrices:

> ***Theorem 7:*** Suppose A and B are matrices where B is obtained from A by applying a single elementary row operation to A. Then $B = E\,A$, where E is the elementary matrix obtained by applying the same elementary row operation to the identity matrix.

In other words, we can carry out any elementary row operation simply by multiplying by the corresponding elementary matrix. Consequently, we can take any matrix A and compute its reduced row echelon form by a sequence of matrix multiplications. Example 1D will step you through an example of how this can be done.

Row Reduction and Matrix Inverses

==================

Example 1D: We will reduce the matrix A defined below to reduced row echelon form, and we will carry out the reduction by multiplying A on the left by the elementary matrices corresponding to the row operations we wish to perform on A.

```
> A := matrix([[2,4],[1,0]]);
> rref(A);
>
```

The command rref(A) shows us that the reduced row echelon form of A is the identity matrix I. So we want to reduce A to I by using elementary matrices.

Step 1: Swap row 1 and row 2. We multiply by the corresponding elementary matrix E_1.

```
> I2 := identmat(2):
> E1 := swaprow(I2,1,2);
> A1 := evalm(E1&*A);
```

Step 2: Add -2 times row 1 to row 2. We multiply by the corresponding elementary matrix E_2.

```
> E2 := addrow(I2,1,2,-2);
> A2 := evalm(E2&*A1);
```

Step 3: Multiply row 2 by 1/4. We multiply by the corresponding elementary matrix E_3.

```
> E3 := mulrow(I2,2,1/4);
> A3 := evalm(E3&*A2);
>
```

We have found the reduced row echelon form of the matrix A by multiplying by three elementary matrices. In fact, we have done much more. We have found the inverse of matrix A! Here's why: Let C denote the product $E_3\, E_2\, E_1$. Then the computation above shows that $C\,A = I$. We must also show that $A\,C = I$. But we know that C is an invertible matrix, since C is the product of invertible matrices. Therefore we can multiply the equation $C\,A = I$ on the left by $C^{(-1)}$ and simultaneously on the right by C and then simplify:

$$C\,A = I$$
$$C^{(-1)}\,C\,A\,C = C^{(-1)}\,I\,C$$
$$A\,C = I$$

The following Maple calculations confirm that C is indeed the inverse of A:

```
> C := evalm(E3&*E2&*E1);
> evalm(C&*A);
  evalm(A&*C);
>
```

==================

Exercise 1.4: If we reduce A to the identity matrix using a different sequence of row operations, would we get a different inverse matrix? No, we proved in Module 5 that a matrix has only one inverse. Check this assertion by reducing A to the identity matrix again, but this time start with the elementary matrices F_1 and F_2 below. Then use this new collection of elementary matrices to compute $A^{(-1)}$.

```
> A := matrix([[2,4],[1,0]]);
  I2 := identmat(2):
> F1 := mulrow(I2,1,1/2);
  A1 := evalm(F1&*A);
> F2 := addrow(I2,1,2,-1);
> A2 := evalm(F2&*A1);
>
```

Student Workspace

Answer 1.4

Summary: If we can reduce A to the identity matrix via a sequence of elementary row operations, then A has an inverse. The inverse is simply the product of the elementary matrices that correspond to the elementary row operations we perform on A. We highlight this fact:

> ***Theorem 8:*** If a matrix A is row equivalent to the identity matrix, then A is invertible.

Exercise 1.5: Check that the matrix A below is not invertible and therefore not row equivalent to the identity matrix.

```
> A := matrix([[4,3,6],[8,6,12],[4,5,3]]);
>
```

Student Workspace

Answer 1.5

Section 2. Invertibility Theorem Revisited

We recall the Invertibility Theorem (IT) from Section 2 of Module 5:

> ***Theorem 6:*** Suppose A is an n by n matrix. Then the statement that A is invertible is equivalent to each of the following statements:
> (a) The only solution to the homogeneous equation $A\,x = 0$ is the trivial solution $x = 0$.
> (b) The columns of A are linearly independent.
> (c) The equation $A\,x = b$ has a solution for every vector b in R^n.
> (d) The columns of A span R^n.

```
>
```

In Module 5, we proved that if A is invertible, then A does have all four of the properties (a), (b), (c), (d). Now, with the help of Theorem 8 above, we can complete the proof:

Proof that Properties (a), (b), (c), or (d) Imply Invertibility

(a) Suppose we are given that the only solution to the homogeneous equation $A\,x = 0$ is the trivial solution $x = 0$. From this we conclude that A is row equivalent to I. (Otherwise, A would have fewer than n pivots and therefore the solution set of $A\,x = 0$ would have a free variable.) By Theorem 8, since A is row equivalent to I, the matrix A is invertible.

(b) It's not hard to see that statements (a) and (b) are equivalent, so we'll skip this proof.

(c) Suppose we are given that the equation $A\,x = b$ has a solution for every vector b in R^n. In particular, we can solve $A\,x = b$ when b is any of the column vectors $e_1, e_2, \ldots, e_n$ of the identity matrix I. Let the corresponding solutions of $A\,x = b$, for each of these columns of I, be denoted by $x_1, x_2, \ldots, x_n$, respectively. Therefore, if C is the matrix whose columns are $x_1, x_2, \ldots, x_n$, then $A\,C = I$:

$$A\,C = A\,[x_1, \ldots, x_n] = [A\,x_1, \ldots, A\,x_n] = [e_1, \ldots, e_n] = I$$

From the fact that $A\,C = I$, it can be shown that A is invertible (see Problem 7).

(d) It's not hard to see that statements (c) and (d) are equivalent, so we'll skip this proof.

```
>
```

Exercise 2.1: Determine whether the matrix A below is invertible or not by determining whether its columns satisfy a linear dependence relation. If the columns do satisfy a linear dependence relation, find one.

```
> A := matrix([[-1,-1,-1,-4],[2,1,0,3],[2,1,-2,-1],[1,1,0,2]]);
>
```

Student Workspace

Answer 2.1

Problems

Problem 1: Invertible matrices and products of elementary matrices

(a) Reduce the matrix A below to the identity matrix, and find elementary matrices E_1, E_2, E_3, E_4 such that $E_4\,E_3\,E_2\,E_1\,A = I$.

(b) (By hand) Express $A^{(-1)}$ as a product of elementary matrices.

(c) (By hand) Express A as a product of elementary matrices. Hint: Solve for A in the equation $E_4\,E_3\,E_2\,E_1\,A = I$.

(d) Based on your work above, describe a general method for writing every invertible matrix as a product of elementary matrices.

```
> A := matrix([[-2,4],[6,-8]]);
>
```

Student Workspace

Problem 2: Elementary matrices perform row operations

Recall that the vector-matrix product $w\,B$ can be written as a linear combination of the rows of B, where the weights are the components of w. Also, each row of the matrix-matrix product $A\,B$ is the product of the corresponding row of A multiplied by all of B. Thus, combining these two ideas, we see that each row of $A\,B$ is a linear combination of the rows of B where the weights are the entries of the corresponding row of A. For example, if A and B are 2 by 2 matrices,

$$A = \begin{bmatrix} r & s \\ t & u \end{bmatrix} \quad B = \begin{bmatrix} a & b \\ c & d \end{bmatrix}$$

then

$$A\,B = \begin{bmatrix} r\,[a, b] + s\,[c, d] \\ t\,[a, b] + u\,[c, d] \end{bmatrix}$$

[>

(a) (By hand) For the matrices E and C below, write the product $E\,C$ in the above form (i.e., each row of the product is a linear combination of the rows of C.

(b) Explain why the product $E\,A$, where E is any n by n elementary matrix of the addrow(..) type and A is any n by m matrix, is exactly the matrix we would get by applying the same row operation to A as the row operation that created E. Hint: Your explanation should be based on the results you observed in part (a).

```
> C := matrix([[a,b],[c,d],[e,f]]);
> I3 := identmat(3):
> E := addrow(I3,1,3,-4);
>
```

Student Workspace

Problem 3: If A is row equivalent to I, Then A is invertible (an essay)

Explain clearly and carefully how we justify the following central fact developed in this module: If a matrix A is row equivalent to the identity matrix, then the matrix A is invertible. Your explanation should be an extrapolation of Example 1D. That is, your explanation should apply to every square matrix that is row equivalent to the identity matrix, not just the specific matrix in Example 1D.

[>

Student Workspace

Problem 4: Invertible or noninvertible at a glance

For each of the matrices below, determine whether or not the matrix is invertible by inspection (i.e., doing no calculation). For each matrix, base your determination on one of the four properties of the Invertibility Theorem, and explain how you know that the matrix does or does not have that property.

[>

(a) $A = \begin{bmatrix} 2 & 4 & -7 \\ 0 & 0 & 0 \\ 8 & -3 & 5 \end{bmatrix}$

(b) $A = \begin{bmatrix} 1 & -3 & 2 \\ 2 & 0 & 1 \\ 0 & 4 & -2 \end{bmatrix}$

(c) $A = \begin{bmatrix} 0 & 1 & 0 & 0 \\ 0 & 0 & 1 & 0 \\ 0 & 0 & 0 & 1 \\ 1 & 0 & 0 & 0 \end{bmatrix}$

(d) A is a square matrix, and there is a nonzero vector x such that $A^T x = 0$.

Student Workspace

Problem 5: Rows and columns

(a) For each of the matrices below, check that the rows are linearly independent if and only if the columns are.
(b) Explain why, for every square matrix A, the rows of A will be linearly independent if and only if the columns are. Hint: Consider the transpose of A.

```
> A := matrix([[1,3],[4,-1]]);
> B := matrix([[1,3],[-2,-6]]);
> C := matrix([[0, 3, -4], [-5, -2, -1], [-9, 5, 8]]);
> E := matrix([[-2,0,-2,-1],[2,1,1,2],[6,4,2,5],[-3,-5,2,0]]);
>
```

Student Workspace

Problem 6: Hand method for finding inverses

Execute the four commands below, which illustrate a common method for computing the inverse of a matrix by hand.

```
> A := randmat(4,4);
> B := augment(A,identmat(4));
> rref(B);
> inverse(A);
>
```

The steps of this method for finding $A^{(-1)}$ are: (1) form the augmented matrix $[A, I]$, where I is the identity matrix; (2) reduce $[A, I]$ to reduced row echelon form. Then $A^{(-1)}$ appears where I had been.
(a) Apply this method by hand to the matrix B below. (You may use Maple to check your answer.)

$$B = \begin{bmatrix} 2 & 2 & 1 \\ -2 & -2 & 0 \\ -1 & 0 & 0 \end{bmatrix}$$

Explanation of why this method works: If you were to solve the equation $B\,x_1 = e_1$, where e_1 is the first column of I, the solution x_1 would be the first column of $B^{(-1)}$. The other columns of $B^{(-1)}$ are found similarly. By reducing the matrix $[B, I]$ to reduced row echelon form, we are simply solving all of the equations $B\,x_1 = e_1$, $B\,x_2 = e_2$, ... simultaneously.

(b) (By hand) Solve the equation $B\,x = e_1$ for the matrix B above, and check that the solution x is the first column of $B^{(-1)}$.

Student Workspace

Problem 7: If $A\,C = I$ then $C\,A = I$

Recall the definition of an invertible matrix: An n by n matrix A is invertible if there is an n by n matrix C such that both $A\,C = I$ and $C\,A = I$. In fact, only one of these two equations is needed, according to Theorem 9:

> ***Theorem 9:*** Suppose A and C are n by n matrices such that $A\,C = I$. Then A and C are both invertible, and $A^{(-1)} = C$.

Prove Theorem 9. Hint: First prove that C is invertible by using the equation $A\,C = I$ to show that C satisfies property (a) of the Invertibility Theorem.

Student Workspace

Pencil and Paper Practice Questions

Note: Maple is not required for any of the problems in this section. Do all of these by hand.

1. Find elementary matrices E_1, E_2, ..., that reduce the matrix A to the matrix B (see below). Use these elementary matrices to find an invertible matrix C such that $C\,A = B$.

$$A = \begin{bmatrix} -1 & 4 \\ 3 & -6 \end{bmatrix} \quad B = \begin{bmatrix} -1 & 4 \\ 0 & 1 \end{bmatrix}$$

2. Suppose that an n by n matrix A has been reduced to row echelon form but not reduced row echelon form. How can you tell by inspection of the row echelon form whether A is invertible or not?

3. If a square matrix A is not row equivalent to the identity matrix, explain why A is not invertible. Hint: Use part (a) of the Invertibility Theorem.

4. Suppose M is an n by n matrix and F, G, H are n by n elementary matrices such that $F\,G\,H\,M = I$. Express M and $M^{(-1)}$ in terms of F, G, and H.

Linear Algebra Modules Project
Chapter 4, Module 1

Geometry of Matrix Transformations of the Plane

Purpose of this module

The purpose of this module is to introduce the basic geometric matrix transformations of the plane and to study their effects on figures in the plane. From this study of the geometric behavior of matrix transformations, you will gain insights into topics studied earlier (such as matrix multiplication) and topics to be studied later (such as eigenvectors).

Prerequisites

Matrix multiplication; inverse of a matrix.

Commands used in this module

```
> restart: with(linalg): with(plots): with(lamp):
>
```

Tutorial

Section 1: Introduction to Matrix Functions

One way of understanding the meaning of multiplying a vector x by a matrix A to obtain a vector $y = A\,x$ is to think of the vector y as a function of the vector x:

$$y = \mathrm{f}(x) = A\,x$$

We might describe this matrix function in words as "multiply by A"; that is, to find the "output" vector y from the "input" vector x, we multiply the input by A.

It is common practice to refer to these matrix functions as *matrix transformations*. Rather than use the letter f as the name of such a function, we will normally use the capital letter T (for transformation). Thus, we will write $y = \mathrm{T}(x)$ rather than $y = \mathrm{f}(x)$. We will also say that the transformation T "maps" an input vector x to the output vector $y = \mathrm{T}(x)$ or simply "T maps x to y."

In this module you will investigate five types of geometric transformations: dilation, shear, rotation, reflection and projection. Each is defined by a particular type of 2 by 2 matrix.

```
>
```

Section 2: Dilation Transformations

The standard form for the dilation matrix is shown below.

$$\begin{bmatrix} r & 0 \\ 0 & r \end{bmatrix}$$

We use the command diag(r,r) to construct a dilation matrix with dilation factor r. For example:

```
> A := diag(.75,.75);
```

We apply this dilation transformation to a vector by multiplying the vector on the left by A. For example, here is the calculation of $A\,u$ when $u = \begin{bmatrix} 1 \\ 2 \end{bmatrix}$:

```
> u := colvector([1,2]);
  Au := evalm(A&*u);
>
```

Alternatively, we can use function notation. Let T be the function that takes as input the vector x and returns as output the vector $A\,x$. Note that the domain for this function is the set of all vectors in R^2. Symbolically, we write $\mathrm{T}(x) = A\,x$. We can define this function in Maple as shown on the next line.

```
> T := x->evalm(A&*x);
```

Now we can find $A\,u$ again, this time by applying the function T to u:

```
> T(u);
>
```

So T maps $\begin{bmatrix} 1 \\ 2 \end{bmatrix}$ to $\begin{bmatrix} .75 \\ 1.5 \end{bmatrix}$.

The geometric effect of applying this transformation is very simple. T simply scales a vector by the factor r. When $r > 1$ the vector is stretched (dilated), and when $0 < r < 1$ the vector is contracted. What does T do when $r < 0$?

We can see the effect of the dilation transformation by looking at the relationship between its input and output vectors. Execute the next input region to see the effect of the transformation $\mathrm{T}(x) = A\,x$ on two vectors u and v.

Note: The input vectors u and v are colored red and their corresponding output vectors $A\,u$ and $A\,v$ are blue. Throughout the module we will follow this color scheme: inputs red and outputs blue.

```
> u := colvector([1,2]);
  v := colvector([3,1]);
> pict1 := drawvec2d(u,v):
  pict2 := drawvec2d(T(u),T(v),headcolor=blue):
  display([pict1,pict2],view=[-2..4,-2..4]);
>
```

Exercise 2.1: (By hand) Identify the vector v in the plot above. Calculate the coordinates of the vector $\mathrm{T}(v)$ without using Maple. Check that your answer is consistent with what you see in the plot.

Student Workspace

Answer 2.1

[>

Applying the same transformation to a larger collection of input vectors gives us a picture that helps us visualize how this transformation contracts vectors towards the origin:

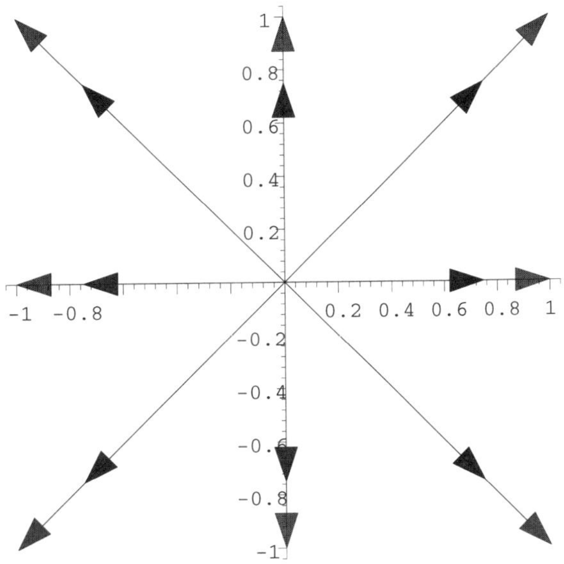

[>

Transforming Figures

Often we want to think of a matrix transformation as acting on points rather than vectors. From this perspective, the transformation $T(x) = A\,x$ moves the point at the head of vector x to the point at the head of vector $A\,x$. Now we can imagine the transformation being applied to arbitrarily complicated figures, since any figure can be thought of as a collection of points. Furthermore, a matrix transformation T will map the line segment from point P to point Q to the line segment from $T(P)$ to $T(Q)$ (unless $T(P) = T(Q)$, in which case the line segment maps to a single point). Therefore it will be convenient to describe figures by a sequence of points connected by straight line segments.

Here's an example involving a figure described by a predefined matrix called "house." Execute the next input region to see the entries in this matrix and the figure it describes.

```
[ > H := evalm(house);
[ > drawmatrix(H);
[ >
```

The command drawmatrix(..) works as follows. Suppose you have in mind a figure made up of n points connected by line segments. If you store these points as the columns of a 2 by n matrix M, then drawmatrix(M) will draw your figure by connecting consecutive points with line segments. To connect the last point back to the first point, we include a last column equal to the first column. You can also change the color of the figure from the default color red to any other color by using drawmatrix(M,figcolor=colorname).

===================

Example 2A: In the input region below, we define a dilation matrix A with $r = 3$ and then compute the matrix product $A\ H$. Compare the result to the original "house" matrix H. Notice that each column of $A\ H$, the output matrix, is the result of multiplying the corresponding column of the input matrix H on the left by A.

```
[ > A := diag(3,3);
[ > H := evalm(house);
[ > AH := evalm(A&*H);
```

Now to see the input (in red) and output (in blue) together, we use the display command:

```
[ > part1 := drawmatrix(H):
  > part2 := drawmatrix(AH,figcolor=blue):
  > display([part1,part2]);
[ >
```

===================

We can accomplish all of this in one step by using the command transform(M,F). Here M is the matrix for the transformation and F is the matrix of points that defines the figure. The result is a plot of both the original figure and the transformed figure. Execute the next line to try it out. Both drawmatrix(..) and transform(..) will be used throughout the rest of the module.

```
[ > transform(A,H);
[ >
```

Exercise 2.2:

(a) Define your own figure as follows: Pick three points from quadrants I and II, and form the 2 by 4 matrix that has these points as its columns (the last column being a repeat of the first). Call this matrix "triangle".

(b) Apply the drawmatrix(..) command to your figure to see what it looks like.

(c) Apply the dilation matrix A to your figure by using the transform(..) command. What are the coordinates of the vertices of the transformed triangle?

[+] Student Workspace

[+] Answer 2.2

Scaling Transformations

As you have observed, a dilation matrix expands (or contracts) vectors by an equal factor in the x and y directions. It is therefore a special case of a more general scaling transformation which scales in the x and y directions by different factors. This takes the familiar form of a diagonal matrix:

$$\begin{bmatrix} r & 0 \\ 0 & s \end{bmatrix}$$

====================

Example 2B: The command diag(r,s) constructs a scaling matrix for given values of r and s. Here we define a scaling matrix A and apply it to the "house" matrix:

```
[ > A := diag(.4,2);
[ > transform(A,house);
[ >
```

====================

We have been concentrating on the effect of matrix transformations on various figures, but of course a transformation maps every point in the plane to an image point. To visualize this movement of all the points of the plane, think of the transformation as distorting the coordinate grid of the plane. We have predefined a "grid" figure that represents a portion of the first quadrant:

```
[ > drawmatrix(grid);
[ >
```

Execute the next input region to see how the scaling matrix A transforms this part of the plane.

```
[ > transform(A,grid);
[ >
```

Try to predict how the grid will be transformed by the following scaling matrix.

```
[ > A := diag(3,-.5);
[ > transform(A,grid);
[ >
```

Now that we have established a quick way of visualizing the impact of transformations, we can move on to some more interesting transformations and figures.

Section 3: Shear Transformations

The standard matrices for the horizontal and vertical shear transformations are shown below.

$$\text{horizontal shear} = \begin{bmatrix} 1 & k \\ 0 & 1 \end{bmatrix} \qquad \text{vertical shear} = \begin{bmatrix} 1 & 0 \\ k & 1 \end{bmatrix}$$

The commands xshearmat(k) and yshearmat(k) construct these matrices with the parameter k. Execute the next input region to see how the vertical shear matrix A transforms the grid figure.

```
[ > A := yshearmat(2);
[ > transform(A,grid);
[ >
```

Observe the following about this picture. The points on the y axis are mapped to themselves. (We say these are "fixed points" of the mapping.) Each of the other vertical lines in the grid is mapped to itself, although the individual points of the line are moved up. Each horizontal line to sheared to a line with slope 2, where 2 is the value of the parameter k in the shear matrix. In the next example, we verify some of these observations for all vertical shear transformations.

====================

Example 3A: Show that a vertical shear matrix with parameter k has the following properties:

(a) The image of each horizontal line $y = b$ is a line with slope k and y intercept b.

(b) The points on the y axis are "fixed points"; that is, they are not moved by the shear.

Solution: (a) We can represent a horizontal line $y = b$ parametrically by $u = \begin{bmatrix} t \\ b \end{bmatrix}$, where t is a parameter. Multiplying this vector by a general vertical shear matrix gives us a parametric description of the corresponding image vectors $A\,u$:

```
[ > u := colvector([t,b]);
[ > A := yshearmat(k);
[ > Au := evalm(A&*u);
[ >
```

Note that $A\,u$ describes a line with slope k and y intercept b.

(b) Multiply A by an arbitrary point $v = \begin{bmatrix} 0 \\ b \end{bmatrix}$ on the y axis. Note that $A\,v = v$, which means that each point on the y axis is a fixed point.

```
[ > v := colvector([0,b]);
[ > Av := evalm(A&*v);
[ >
```

====================

Similarly, a horizontal shear matrix with parameter k maps each vertical line $x = a$ to a line with slope $1/k$ and x intercept a:

```
[ > transform(xshearmat(2),grid);
[ >
```

Exercise 3.1: Show that a horizontal shear matrix has the following properties:
(a) The points on the x axis are fixed points.
(b) The image of each horizontal line is the same horizontal line.

Student Workspace

Answer 3.1

Section 4: Composite Transformations

If A and B are two matrix transformations of the plane, their product $A\,B$ is another matrix transformation. The matrix $A\,B$ is referred to as a "composite" transformation for the following reason. If T_1 and T_2 are the matrix transformations defined by $T_1(x) = A\,x$ and $T_2(x) = B\,x$, then the composite function $T_1(T_2(x))$ equals $A\,B\,x$:

$$T_1(T_2(x)) = A\,(B\,x) = (A\,B)\,x \quad \text{and likewise} \quad T_2(T_1(x)) = B\,(A\,x) = (B\,A)\,x$$

```
[ >
```

In general, we would not expect the composite transformations $T_1(T_2(x))$ and $T_2(T_1(x))$ to have the same effect, just as we wouldn't in general expect the matrix products $A\,B$ and $B\,A$ to be equal. However, if two matrix transformations commute (i.e., $A\,B = B\,A$), we may be able to see this by visualizing how these transformations behave geometrically. For example, based on what you know about dilations and shears, would you predict that these transformations commute with each other?

Exercise 4.1: Let A be a dilation matrix with $r = 2$, and let B be a horizontal shear with $k = 3$:

```
> A := diag(2,2);
  B := xshearmat(3);
>
```

(a) Apply the matrix transformations $A\,B$ and $B\,A$ to the house figure. Verify that they yield the same result.

(b) Show that for every dilation matrix A and every shear matrix B, we have $A\,B = B\,A$.

Student Workspace

Answer 4.1

Section 5: Rotation Transformations

====================

Example 5A: Find the matrix transformation R that rotates every vector $\begin{bmatrix} a \\ b \end{bmatrix}$ counter-clockwise 90 degrees. For example, the matrix R transforms the vector $\begin{bmatrix} 3 \\ 2 \end{bmatrix}$ to the vector $\begin{bmatrix} -2 \\ 3 \end{bmatrix}$ as shown below:

```
>
```

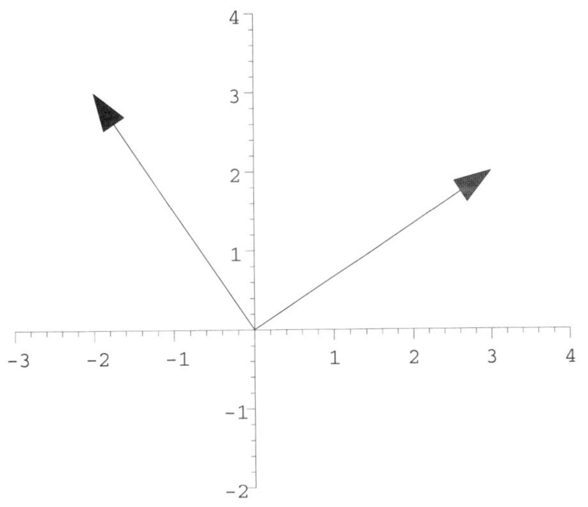


```
>
```

Solution: Just as in the example above, R transforms a general vector $\begin{bmatrix} a \\ b \end{bmatrix}$ to the vector $\begin{bmatrix} -b \\ a \end{bmatrix}$:

$$R \begin{bmatrix} a \\ b \end{bmatrix} = \begin{bmatrix} -b \\ a \end{bmatrix}$$

We can now find R by using the method of decomposition:

$$\begin{bmatrix} -b \\ a \end{bmatrix} = a\begin{bmatrix} 0 \\ 1 \end{bmatrix} + b\begin{bmatrix} -1 \\ 0 \end{bmatrix} = \begin{bmatrix} 0 & -1 \\ 1 & 0 \end{bmatrix}\begin{bmatrix} a \\ b \end{bmatrix}$$

Therefore $R = \begin{bmatrix} 0 & -1 \\ 1 & 0 \end{bmatrix}$.

```
[ >
```

====================

More generally, the matrix transformation that rotates every vector in the plane by θ radians in the counter-clockwise direction is:

$$\begin{bmatrix} \cos(\theta) & -\sin(\theta) \\ \sin(\theta) & \cos(\theta) \end{bmatrix}$$

The rotatemat(θ) command constructs a rotation matrix for any angle θ. For example, we construct a rotation matrix A corresponding to a rotation angle of π/6 radians (30 degrees) and apply it to the "grid" figure:

```
[ > A := rotatemat(Pi/6);
[ > transform(A,grid);
[ >
```

Next let B be the rotation matrix corresponding to an angle of π/3 radians (60 degrees):

```
[ > B := rotatemat(Pi/3);
[ >
```

Exercise 5.1: (a) Check that the matrices A and B are related by the equation $A^2 = B$, and give a geometric explanation for this relationship.

(b) Find a positive integer k such that $A^k = I$.

[+] Student Workspace

[+] Answer 5.1

In the next input region we construct a new rotation matrix U and apply it to a new figure called "bug":

```
[ > U := rotatemat(Pi/4);
[ > transform(U,bug);
[ >
```

Let's find the inverse of U using a geometric approach.

Exercise 5.2:

Finding an inverse of U is equivalent to finding a second rotation transformation V that returns the bug to its original position (since $V\,U = I$ and I maps all points to themselves).

(a) Do this two ways, one where V rotates vectors clockwise and one where V rotates vectors counterclockwise. Check that both ways of finding the inverse of U give the same result as the command inverse(U).

(b) Express the inverse of U as a positive power of U; i.e., find a positive integer k such that $V = U^k$.

Student Workspace

Answer 5.2

Section 6: Using Transformations to Animate Figures

If we take a rotation matrix A with a small angle such as $\pi/18$ and apply it to the house figure H, we will get a slightly turned house $A\ H$. If we apply the matrix A again to the slightly turned house, we will turn it a bit more. Now we have $A\ A\ H$ or $A^2\ H$. If we keep doing this, we will produce a succession of still pictures which can be used to create an animation of a rotating house.

====================

Example 6A: The command movie(..) displays just such a sequence of still pictures that constitute an animation. The syntax is movie(A,H,n), where A is the matrix that you want to apply repeatedly, H is the matrix describing the figure, and n is the number of applications (i.e., iterations) of the matrix A. Keep in mind that the frames consist of A, $A\ H$, $A^2\ H$, ..., $A^n\ H$. Execute the next input region for an example. (This may take some time to compute.) Then click on the picture to bring up the animation buttons, and click on the "play" button.

```
> H := evalm(house);
  A := rotatemat(Pi/18);
> movie(A,H,8);
>
```

To see the succession of frames that make up the animation, add the option frames=on:

```
> movie(A,H,8,frames=on);
>
```

====================

====================

Example 6B: If we form a composite transformation made up of a contraction and a rotation, we can produce a more dynamic movie . We will use a slight contraction followed by the small rotation above.

```
> C := diag(.95,.95);
> evalm(A);
> M := evalm(A&*C);
> movie(M,H,8);
>
```

Here are the frames for that animation.

```
> movie(M,H,8,frames=on);
>
```

====================

Feel free to experiment. Just keep in mind that the computing time will increase with the complexity of the figure and the number of iterations. In a later module on computer graphics, we will come back to this topic.

Section 7: Reflection Transformations

===================

Example 7A: Find the matrix transformation R that reflects every vector $\begin{bmatrix} a \\ b \end{bmatrix}$ across the x axis. For example, the matrix R transforms the vector $\begin{bmatrix} 3 \\ 2 \end{bmatrix}$ to the vector $\begin{bmatrix} 3 \\ -2 \end{bmatrix}$ as shown below:

```
>
```

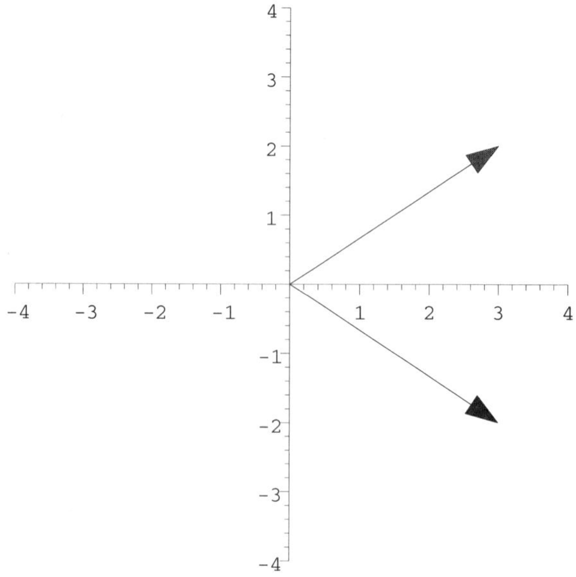


```
>
```

Solution: R simply changes the sign of the second coordinate, transforming $\begin{bmatrix} a \\ b \end{bmatrix}$ to $\begin{bmatrix} a \\ -b \end{bmatrix}$. We use the method of decomposition to find the matrix transformation R:

$$R\begin{bmatrix} a \\ b \end{bmatrix} = \begin{bmatrix} a \\ -b \end{bmatrix} = a\begin{bmatrix} 1 \\ 0 \end{bmatrix} + b\begin{bmatrix} 0 \\ -1 \end{bmatrix} = \begin{bmatrix} 1 & 0 \\ 0 & -1 \end{bmatrix}\begin{bmatrix} a \\ b \end{bmatrix}$$

Therefore R is given by:

```
> R := matrix([[1,0],[0,-1]]);
```

Let's check this matrix transformation by applying it to the bug figure:

```
> transform(R,bug);
>
```

===================

Exercise 7.1: Use the method of decomposition to find the following matrices. Start by considering how each transformation maps a particular vector, such as $\begin{bmatrix} 3 \\ 2 \end{bmatrix}$.

(a) Find the matrix that reflects vectors across the y axis. Check your answer as we did above.

(b) Find the matrix that reflects vectors across the 45 degree line, $y = x$. Check your answer as above.

Student Workspace

Answer 7.1

```
>
```

More generally, the matrix transformation below reflects vectors across the line through the origin making an angle of θ radians with the x axis.

$$\begin{bmatrix} \cos(2\,\theta) & \sin(2\,\theta) \\ \sin(2\,\theta) & -\cos(2\,\theta) \end{bmatrix}$$

The command reflectmat(θ) constructs a reflection matrix for an angle of θ radians. We check that this formula gives the three reflection matrices we have already considered : reflection about the x axis, reflection about the y axis, and reflection about the line $y = x$.

```
> reflectmat(0);
> reflectmat(Pi/2);
> reflectmat(Pi/4);
>
```

When we apply a reflection transformation to a figure, we get its mirror image across the line of reflection. The input region below displays the result of applying a reflection matrix for the angle $\theta = \pi/3$ to the bug. It also displays the line of reflection.

Explore~~~ Change the value of θ below to view the effect of other reflection matrices. Find a value of θ for which the image of the bug lies entirely within quadrant I.

```
> theta := Pi/3;
  R := reflectmat(theta):
  pict1 := transform(R,bug):
  pict2 :=
  plot([t*cos(theta),t*sin(theta),t=-10..40],color=black):
  display([pict1,pict2]);
>
```

Section 8. Projection Transformations

Exercise 8.1: Use the method of decomposition to find the matrix transformation P that projects vectors perpendicularly onto the x axis. That is, P maps $\begin{bmatrix} a \\ b \end{bmatrix}$ to $\begin{bmatrix} a \\ 0 \end{bmatrix}$.

Student Workspace

Answer 8.1

```
>
```

The matrix below is the transformation that projects vectors onto the line through the origin making an angle of θ radians with the x axis.

$$\begin{bmatrix} \cos(\theta)^2 & \cos(\theta)\sin(\theta) \\ \cos(\theta)\sin(\theta) & \sin(\theta)^2 \end{bmatrix}$$

Alternatively, if we apply double-angle identities, we get :

$$\begin{bmatrix} \frac{\cos(2\,\theta)+1}{2} & \frac{\sin(2\,\theta)}{2} \\ \frac{\sin(2\,\theta)}{2} & \frac{1-\cos(2\,\theta)}{2} \end{bmatrix}$$

```
[ >
```

The command projectmat(θ) constructs this projection matrix for an angle of θ radians:

```
> theta := 'theta';
> projectmat(theta);
> 
```

====================

Example 8A: The plot below displays the projections of three vectors u, v, w onto the line making an angle of 30 degrees with the x axis. If we were to drop a perpendicular from the head of the vector u to the line of projection, the foot of the perpendicular would land at the head of $A\ u$; the same is true for v and $A\ v$, and for w and $A\ w$.This is what we mean by a projection; we project vectors onto the line of projection by projecting in the direction perpendicular to that of the line of projection.

On a paper copy of the plot, draw the three perpendiculars, and label the vectors $A\ u$, $A\ v$, $A\ w$.

```
[ >
```

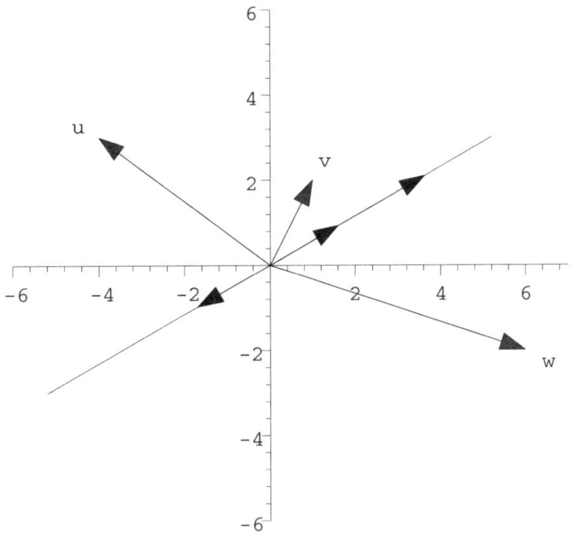

====================

[>

Explore~~~ In the next input region, we project the bug figure onto a 45 degree line. To change the line of projection, change the value of θ.

```
> theta := Pi/4;
  A := projectmat(theta):
  p1 := transform(A,bug):
  p2 := plot([t*cos(theta),t*sin(theta),t=-5..40],color=black):
  display([p1,p2]);
```

[>

Exercise 8.2: (By hand) (a) What are the fixed points for a projection transformation? (Recall that a fixed point is one that is mapped to itself by the transformation.)

(b) Which points are mapped to the zero vector; that is, for which x is $T(x) = 0$?

Student Workspace

Answer 8.2

[>

Problems

Problem 1: Guess the transformation

This problem consists of three questions. In each, you are presented with a picture that shows the effect of a transformation on the bug. (The input is red, the output green.) Your task is to identify the specific transformation (or composite transformation) that is being applied.There may be more than one answer. List two if possible.

Execute the next input region to see Question 1.

```
> display(question(1));
>
```

The next input region provides a quick way for you to check your answers. For example, suppose your answer to Question 1 is the rotation matrix *A* below. The display(..) command lets you see the output from *A* (in blue) superimposed on the question plot.

```
> A := rotatemat(Pi/2);
> your_answer := drawmatrix(A&*bug,figcolor=blue):
> display([your_answer,question(1)]);
>
```

Execute the next input region to see Question 2.

```
> display(question(2));
>
```

Execute the next input region to see Question 3.

```
> display(question(3));
>
```

Problem 2: Easy inverses

In Exercise 5.2 you found an easy way to state the inverse of a rotation matrix. Is there an easy way to state the inverse for each of the other types of transformations (i.e., dilations, shears, reflections, and projections)? If so, explain, from a geometric viewpoint, how the inverse of each type of transformation can be found. You may use Maple to check your answers.

Student Workspace

Problem 3: Line segments transform to line segments

The line segment from a point p to another point q is expressed parametrically as $p + t\,(q - p)$, with the parameter t ranging from $t = 0$ to $t = 1$. Check that when $t = 0$ we are at p and when $t = 1$ we are at q. If we apply a matrix transformation A to the points on this line segment, we get:

$$A\,(p + t\,(q - p)) = A\,p + t\,(A\,q - A\,p) \qquad [1]$$

Note that this is just the line segment between the points $A\,p$ and $A\,q$. If $A\,p = A\,q$, then the image segment collapses to the single point $A\,p$.

```
>
```

(a) Explain how we get from the left to the right side of expression [1] using the rules of matrix algebra.

(b) Let A be the rotation matrix rotatemat(Pi/3), and let $p = \begin{bmatrix} 3 \\ 4 \end{bmatrix}$, $q = \begin{bmatrix} 4 \\ 3 \end{bmatrix}$. Find the parametric description of the line segment from p to q, and then find the parametric description of the image of this line segment when the rotation matrix is applied to it.

(c) Give an example of a nonzero matrix *B* that maps the line segment from p to q in (b) to a single point. Hint: Think geometrically.

Student Workspace

Problem 4: Parallel lines transform to parallel lines

Show that a matrix transformation always takes a pair of parallel lines to parallel lines (unless it sends each line to a point). Hint: Express the two lines in vector form, $p + t\,v$.

Student Workspace

Problem 5: Determinant of a matrix transformation

The matrices that you have worked with in this module are all "well-behaved" in the sense that each has an identifiable geometric impact. But since any matrix will suffice to define a matrix transformation, perhaps we should take a quick look at some of the other possibilities. In the region below, we generate random 2 by 2 matrices and apply them to the "grid" figure. To keep things simple, we consider only matrices that have integer entries ranging between -2 and 2. Execute this region many times to see the variety of possibilities. As you do so, some familiar transformations from this module will occasionally make an appearance. In some cases you may recognize the result as a composite of two (or more) of our basic geometric transformations.

```
>
```

Each time we generate a new matrix we also calculate the absolute value of its determinant. This number tells us something about the relationship between the original (red) grid and its image (blue) grid.

(a) Find the relationship and illustrate it by referring to several examples. If you find an example you want to save for purposes of illustration, just cut and paste it into your discussion.

(b) (Challenge) Give an algebraic proof of the relationship you found in part (a) using properties of determinants.

```
> A := randmat(2,2,bound=2);
  abs(det(A));
> transform(A,grid);
>
```

Student Workspace

Problem 6: True/False questions about reflections

(By hand) These are true/false questions. Give a geometric reason that supports your answer.

(a) If *A* is the reflection matrix for an angle θ, then A^2 is the reflection matrix for an angle 2 θ.

(b) If *A* is the reflection matrix for an angle θ and *B* is the reflection matrix for -θ, then *A* and *B* must be inverse matrices.

(c) If *A* is the reflection matrix for an angle θ, then $A^n = A$, whenever *n* is a positive odd integer.

Student Workspace

Problem 7: Square roots, cube roots, ...

In the real numbers, equations such as $x^2 = 1$ and $x^3 = 1$ and $x^4 = 1$ have only one or two solutions. On the other hand, matrix equations of the same form have many more solutions. Using only our five special types of matrix transformations, find as many solutions as you can to each of the equations $A^2 = I$ and $A^3 = I$ and $A^4 = I$. Explain geometrically why your solutions work.

Student Workspace

Problem 8: Products of projections I

Let A and B be the projection matrices below.

(a) Explain geometrically why $A\,B = B\,A = O$. Hint: Draw v, $B\,v$, and $A\,B\,v$ for a few vectors v.

(b) Calculate A^2 and B^2. What do you observe? Is this true for all projection matrices? Explain geometrically.

```
> A := projectmat(Pi/4);
  B := projectmat(3*Pi/4);
>
```

Student Workspace

Problem 9: Products of projections II

This is a continuation of Problem 8.

(a) (By hand) Define a third projection matrix C (see below) and a vector $v = \begin{bmatrix} 3 \\ 1 \end{bmatrix}$. Make a hand sketch (or use drawvec2d(..)) that shows the following four vectors together in a single plot: $A\,v$, $C\,v$, $A\,C\,v$ and $C\,A\,v$. Use your sketch to explain why the matrix products $A\,C$ and $C\,A$ are not equal. Are there any vectors x where $A\,C\,x$ does equal $C\,A\,x$? Hint: Solve the homogeneous equation $(A\,C - C\,A)\,x = 0$ and thus find all vectors x such that $A\,C\,x = C\,A\,x$.

(b) Find another pair of projection matrices that, like A and B in Problem 8, commute. Under what conditions do two projection matrices commute?

(c) Is it possible to find two nonzero matrices from one of the other types of transformations (i.e., dilations, shears, rotations or reflections) whose product is the zero matrix? Explain why your answers are correct.

```
> A := projectmat(Pi/4);
  B := projectmat(3*Pi/4);
  C := projectmat(Pi/3);
>
```

Student Workspace

Problem 10: Reflections that commute

(a) If P is the matrix that projects onto the line through the origin making an angle of θ radians with the x axis, and R is the matrix that reflects across this line, use trigonometric identities to show that

$R = 2\,P - I.$

(b) Which reflection matrices commute? Hint: In Problem 9, you found the conditions under which projection matrices commute. Use your result from that problem and part (a) of this problem.

Student Workspace

Problem 11: Derivation of rotation, projection, reflection matrices

(a) Let M be the matrix transformation that rotates vectors counter-clockwise by θ radians. It's not hard to see that M sends the vector $\begin{bmatrix} 1 \\ 0 \end{bmatrix}$ to $\begin{bmatrix} \cos(\theta) \\ \sin(\theta) \end{bmatrix}$, and the vector $\begin{bmatrix} 0 \\ 1 \end{bmatrix}$ to $\begin{bmatrix} -\sin(\theta) \\ \cos(\theta) \end{bmatrix}$. Use this information to derive the formula for M that was shown in Section 5.

(b) Let P be the matrix transformation that projects vectors onto the line through the origin making an angle of θ radians with the x axis. Let u be a unit vector lying along this line and v be a unit vector perpendicular to this line (see below). Then $P\,u = u$ and $P\,v = 0$. Use this information to derive the formula for P that was shown in Section 8. (Hint: Find P by solving a matrix equation of the form $P\,A = B$.)

```
> u := colvector([cos(theta),sin(theta)]);
  v := colvector([-sin(theta),cos(theta)]);
>
```

(c) Let R be the matrix transformation that reflects vectors across the line through the origin making an angle of θ radians with the x axis. Use the method in part (b) to derive the formula for R that was shown in Section 7.

Student Workspace

Problem 12: Transformations of lines

Let A be the projection matrix below and B be the rotation matrix below:

```
> A := projectmat(Pi/2);
  B := rotatemat(Pi/4);
>
```

(a) Find the image of the horizontal line $y = 1$ under each of the matrix transformations A, B, $A\,B$, and $B\,A$. (If you can, determine your answers by geometric reasoning alone.)

(b) Find a matrix transformation that maps the horizontal line $y = 1$ onto the vertical line $x = 2$. (This can be done using only transformations of the type studied in this module.)

(c) Can you find a matrix transformation that maps the x axis to the line $y = 1$?

Student Workspace

Pencil and Paper Practice Questions

Note: Maple is not required for any of the problems in this section. Do all of these by hand.

1. The slant line below is the line of projection for a 2 by 2 projection matrix P and the line of reflection for a 2 by 2 reflection matrix R.
(a) For each of the indicated vectors u, v, w, z, draw the corresponding output vectors $P\,u$, $P\,v$, $P\,w$, $P\,z$ and $R\,u$, $R\,v$, $R\,w$, $R\,z$. Label each output vector.

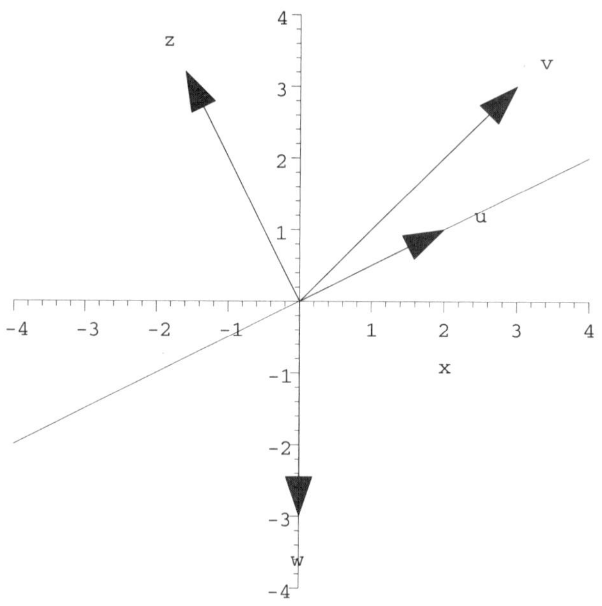

2. If A is a general 2 by 2 rotation matrix with rotation angle θ, what is the transformation A^2? (Do not calculate; just reason geometrically.) Answer the same question if A is a general 2 by 2 projection, reflection, or dilation matrix.

Linear Algebra Modules Project
Chapter 4, Module 2

Geometry of Matrix Transformations of R^3

Purpose of this module

The purpose of this module is to study the basic geometric matrix transformations of R^3 and the concepts of kernel and range of a transformation.

Prerequisites

Chapter 4, Module 1: Geometric Transformations of the Plane.

Commands used in this module

```
> restart: with(linalg): with(plots): with(lamp):
>
```

Tutorial

Section 1: Geometric Tranformations of R^3

As in Module 1, we will study dilations, rotations, reflections, and projections. Now, however, these will be 3 by 3 matrix transformations that map vectors in R^3 to vectors in R^3.

Scaling Transformations

====================

Example 1A: The 3 by 3 matrix A defined below scales the x, y, and z components of vectors in R^3 by different factors. That is, if we mulitply any vector in R^3 by A, the components of the vector will be scaled by the corresponding diagonal entries of A.

```
> A := diag(2,3,5);
> v := colvector([a,b,c]);
  Av := evalm(A&*v);
>
```

We use the drawmatrix3d(..) command to draw a figure in R^3, and the transform3d(..) command to draw both the original figure and the transformed figure. As in Module 1, we have several predefined figures available to use. Execute the next region to see the "hotel3d" figure.

```
> drawmatrix3d(hotel3d);
>
```

In the command below, we apply the scaling transformation A to hotel3d. Note that the dimensions of the original figure (in red) have not changed, but the scale of the plot has changed to accommodate the larger transformed figure (in blue).

```
> transform3d(A,hotel3d);
>
```

====================

The vectors e_1, e_2, e_3 below are called the "standard unit vectors" of R^3. They are vectors of length 1 along the positive x, y, and z axes, respectively. They are also the columns of the 3 by 3 identity matrix.

```
> e1 := colvector([1,0,0]);
  e2 := colvector([0,1,0]);
  e3 := colvector([0,0,1]);
>
```

Since $A\,I = I$ for any 3 by 3 matrix A, the standard unit vectors have the following useful property:

- If A is a 3 by 3 matrix and T is the matrix transformation defined by $\mathrm{T}(x) = A\,x$ for all vectors x in R^3, then the columns of A are $\mathrm{T}(e_1)$, $\mathrm{T}(e_2)$, $\mathrm{T}(e_3)$.

====================

Example 1B: Let B be the matrix below, and let T be the transformation $\mathrm{T}(x) = B\,x$. If we apply T to the standard unit vectors e_1, e_2, e_3 of R^3, then $\mathrm{T}(e_1) = B_1$, $\mathrm{T}(e_2) = B_2$, $\mathrm{T}(e_3) = B_3$, where B_1, B_2, B_3 are the columns of B.

```
> B := matrix([[2,-4,1],[3,0,-2],[-1,1,5]]);
> T := x->evalm(B&*x);
> T(e1);
> T(e2);
> T(e3);
>
```

====================

Exercise 1.1: What are the images of the standard unit vectors under the scaling transformation in Example 1A?

Student Workspace

Answer 1.1

We can use this property of the standard unit vectors in two ways. If we want to construct a matrix transformation, we will be able to do so quickly if we know the images of the standard unit vectors. Alternatively, given a matrix transformation, we may be able to analyze its geometric behavior by interpreting its columns as the images of the standard unit vectors under the transformation.

```
>
```

Rotation Transformations

====================

Example 1C: Let M be the 3 by 3 matrix that rotates vectors 90 degrees counter-clockwise about the z axis. Use the technique in Example 1B to find the entries of M.
Solution: Since the rotation is about the z axis, the transformation T defined by $\mathrm{T}(x) = M\,x$ maps e_1 to e_2, e_2 to $-e_1$, and e_3 to itself. Thus:

$$\mathrm{T}(e_1) = \begin{bmatrix} 0 \\ 1 \\ 0 \end{bmatrix} \quad \mathrm{T}(e_2) = \begin{bmatrix} -1 \\ 0 \\ 0 \end{bmatrix} \quad \mathrm{T}(e_3) = \begin{bmatrix} 0 \\ 0 \\ 1 \end{bmatrix}$$

Therefore the rotation matrix is:

$$M = \begin{bmatrix} 0 & -1 & 0 \\ 1 & 0 & 0 \\ 0 & 0 & 1 \end{bmatrix}$$

We check this by applying the matrix M to the hotel3d figure:

```
[ > M := matrix([[0, -1, 0], [1, 0, 0], [0, 0, 1]]);
[ > transform3d(M,hotel3d);
[ >
```

====================

Exercise 1.2: Use the technique in Example 1B to find the 3 by 3 matrix that rotates vectors 90 degrees counter-clockwise about the x axis. Check your matrix by applying it to hotel3d.

Student Workspace

Answer 1.2

The command rotatemat3d(v,θ) constructs the 3 by 3 rotation matrix that rotates vectors θ radians counter-clockwise about the axis with direction vector v. The two rotation matrices we considered in Examples 1C and Exercise 1.2 are therefore:

```
[ > rotatemat3d([0,0,1],Pi/2);
[ > rotatemat3d([1,0,0],Pi/2);
[ >
```

Reflection Transformations

====================

Example 1D: Let R be the 3 by 3 matrix that reflects vectors in R^3 across the yz plane. Use the method of decomposition from Module 1 to find R.

Solution: Note that $R\begin{bmatrix} a \\ b \\ c \end{bmatrix} = \begin{bmatrix} -a \\ b \\ c \end{bmatrix}$, and then decompose the vector on the right to find R:

$$\begin{bmatrix} -a \\ b \\ c \end{bmatrix} = a\begin{bmatrix} -1 \\ 0 \\ 0 \end{bmatrix} + b\begin{bmatrix} 0 \\ 1 \\ 0 \end{bmatrix} + c\begin{bmatrix} 0 \\ 0 \\ 1 \end{bmatrix} = \begin{bmatrix} -1 & 0 & 0 \\ 0 & 1 & 0 \\ 0 & 0 & 1 \end{bmatrix}\begin{bmatrix} a \\ b \\ c \end{bmatrix}$$

(Alternatively, by the technique in Example 1B, we have $T(e_1) = -e_1$, $T(e_2) = e_2$ and $T(e_3) = e_3$, which yields the same matrix.)

```
[ > R := matrix([[-1, 0, 0], [0, 1, 0], [0, 0, 1]]);
```

Here is the result of the transform3d(..) command applied to the hotel3d figure:

```
[ > transform3d(R,hotel3d);
[ >
```

And here is the same transformation applied to a figure called "jet3d."

```
[ > transform3d(R,jet3d);
[ >
```

====================

Exercise 1.3: (By hand) Find each of the following matrices. Use either the method of decomposition or the technique in Example 1B.

(a) The 3 by 3 matrix A that reflects vectors across the xy plane.

(b) The 3 by 3 matrix B that reflects vectors across the vertical plane $y = x$.

```
[ >
```

⊞ Student Workspace

⊞ Answer 1.3

The command reflectmat3d(n) constructs the 3 by 3 reflection matrix that reflects vectors across the plane through the origin with normal vector n. For example, if we want to reflect across the yz plane (as in Example 1D), we use $n = [1, 0, 0]$.

```
[ > reflectmat3d([1,0,0]);
[ >
```

Projection Transformations

====================

Example 1E: Use the technique in Example 1B to find the 3 by 3 matrix P that projects vectors onto the yz plane.

Solution: The vector e_1 is projected to the zero vector, and e_2 and e_3 are mapped to themselves. So

$$P = \begin{bmatrix} 0 & 0 & 0 \\ 0 & 1 & 0 \\ 0 & 0 & 1 \end{bmatrix}$$

```
[ > P := matrix([[0,0,0],[0,1,0],[0,0,1]]);
```

Here is the transformation applied to hotel3d. If you rotate the figure so you are looking straight at the *yz* plane, you will see the relationship between the 3d figure and its projection.

```
[ > transform3d(P,hotel3d);
[ >
```

The projection of jet3d is also interesting.

```
[ > transform3d(P,jet3d,axes=none);
[ >
```

====================

Exercise 1.4: Find the 3 by 3 matrix P that projects vectors in R^3 onto the vertical plane $y = x$. (Hint: the vectors e_1 and e_2 are projected onto the same vector on the line $y = x$ in the xy plane. A sketch should help).

Student Workspace

Answer 1.4

The command projectmat3d(n) constructs the 3 by 3 projection matrix that projects vectors onto the plane through the origin with normal vector n. For example, if we want to project onto the *yz* plane (as in Example 1E), we use $n = [1, 0, 0]$:

```
[ > projectmat3d([1,0,0]);
[ >
```

Section 2: Kernel and Range

In this section we introduce two concepts associated with matrix transformations that are analogous to concepts you have seen in your study of functions in precalculus and calculus. Consider, for example, $f(x) = x^2 - 4$, which defines a function f from R to R. The set of "zeros" of such a function is the set of values of x such that $f(x) = 0$; in this example, the set of zeros is $\{2, -2\}$. The "range" of such a function is the set of values of $f(x)$, where x ranges over the domain R; in this example, the range is $\{y : -4 \le y\}$.

We now define the analogous concepts for the matrix transformation $T(x) = A\,x$, where A is any m by n matrix. T is thus a transformation that takes vectors in R^n to vectors in R^m.

```
[ >
```

Definition: The ***kernel*** of T is the set of vectors x in R^n such that $T(x) = 0$; in other words, the kernel is the set of "zeros" of T.

Definition: The ***range*** of T is the set of all values of $T(x)$, where x ranges over all vectors in R^m; in other words, the range is the set of all output (or image) vectors of T in R^m.

Exercise 2.1: (By hand) Find the kernel and range for the matrix transformation T that projects vectors in R^3 onto the xz plane. Determine your answers by geometric reasoning.

Student Workspace

Answer 2.1

[>

Both the kernel and range can be described using familiar ideas from our study of linear systems:

- Since $T(x) = A\,x$, the kernel of T is the set of solutions of the homogeneous linear system $A\,x = 0$.
- Since $T(x) = A\,x$, the range of T is the set of all vectors b such that $A\,x = b$ has a solution. That is, a vector b is an image vector for T if and only if there exists some vector x such that $A\,x = b$.

Note that the zero vector is always in both the kernel and range, since $A\,0 = 0$ for every matrix A.

===================

Example 2A: Let T be the matrix transformation that rotates vectors through an angle θ about the z axis. Find the kernel and range of T.

Solution: When you rotate a vector, you cannot get the zero vector unless you start with the zero vector. So the kernel of T consists of just one vector, the zero vector itself, $\{0\}$. The range of T is all of R^3, since every vector v in R^3 is the image of some input vector u in R^3. Specifically, if u is the vector that we get by rotating v back by the same angle θ, we have $T(u) = v$.

[>

===================

Kernel, Range and Invertible Matrices

Many matrix transformations follow the pattern of Example 2A -- kernel = $\{0\}$ and range = R^3. In fact, if the kernel of a 3 by 3 matrix transformation is $\{0\}$ then the range must be R^3! Likewise, if the range is R^3, then the kernel must be $\{0\}$. Furthermore, both of these conditions are equivalent to the matrix being invertible. These facts are contained in Theorem 6 in Chapter 3, the Invertibility Theorem (IT). Let's look again at a precise statement of the relevant parts of the IT:

[>

- If A is a 3 by 3 matrix, the following three statements are equivalent:

- (a) A is invertible.
- (b) The only solution of the homogeneous linear system $A\,x = 0$ is the trivial solution $x = 0$ (i.e., the kernel of the transformation is $\{0\}$).
- (c) The linear system $A\,x = b$ has a solution for every b in R^3 (i.e., the range of the transformation is R^3)

Exercise 2.2: Is the following statement true or false? "No 3 by 3 projection matrix is invertible".

Use a geometric argument based on the discussion above to justify your answer.

Student Workspace

Answer 2.2

[>

Exercise 2.3: In Section 1 we considered projections onto a plane. Now let's project onto a line. Let $T(x) = A\,x$ be the matrix transformation that projects vectors in R^3 onto the vector $\begin{bmatrix} 0 \\ 0 \\ 1 \end{bmatrix}$.

(a) Find the kernel and range of T.
(b) Find the matrix A and apply it to the hotel3d figure.
(c) Does the matrix that you found in (b) have an inverse? Use your answer to (a) to justify your conclusion.

Student Workspace

Answer 2.3

Exercise 2.4: Let T be the projection transformation in Exercise 2.3 and U be the rotation transformation in Exercise 1.2. (U rotates vectors 90 degrees counter-clockwise about the x axis.) Find the kernel and range of the composite transformation $T\,U$. **Caution:** Keep in mind that $T\,U(x) = T(U(x))$, which means that U is applied first to x, then T.

Student Workspace

Answer 2.4

[>

Problems

Problem 1: Reflections in R^3

Let *Rxy*, *Rxz*, *Ryz* be the 3 by 3 matrices that reflect vectors across the *xy*, *xz*, *yz* planes, respectively.
(a) Find each of these matrices by hand. (We found *Ryz* in Example 1D.)
(b) Create a picture that displays the hotel3d figure and its three reflections across the coordinate planes.
(c) Find the 3 by 3 matrix that reflects vectors across the vertical plane $y = -x$. Do this by composing three matrices: a rotation, a reflection, and another rotation, where the reflection is one of the reflections you found in (a). Use the transform3d(..) command to apply this reflection matrix to hotel3d.

Student Workspace

Problem 2: Projections in R^3

Let *Pxy*, *Pxz*, *Pyz* be the 3 by 3 matrices that project vectors onto the the *xy*, *xz*, *yz* planes, respectively.
(a) Find each of these matrices by hand. (We found *Pyz* in Example 1E.)
(b) Create a picture that displays the hotel3d figure and its projections onto the three coordinate planes.
(c) Do *Pxy*, *Pxz*, and *Pyz* commute? That is, are the equations

$$Pxy\ Pxz = Pxz\ Pxy, \quad Pxy\ Pyz = Pyz\ Pxy, \quad Pxz\ Pyz = Pyz\ Pxz$$

true? Give a geometric justification for your answer.

Student Workspace

Problem 3: Projection onto a line in R^3

Let $T(x) = A\,x$ be the transformation that projects vectors onto the line containing the vector $\begin{bmatrix} 1 \\ 0 \\ 1 \end{bmatrix}$.

(a) Find all fixed points of T. (Recall that a "fixed point" of a transformation T is a vector x that is mapped to itself: $T(x) = x$.)
(b) Find the kernel and range of T.
(c) Find the matrix *A*. Check your matrix by applying it to the hotel3d figure.

Student Workspace

Problem 4: Kernel and range of a composite transformation

Let *P* be the 3 by 3 projection matrix that projects vectors onto the *xy* plane and *R* be the 3 by 3 rotation matrix that rotates vectors 90 degrees counter-clockwise about the *y* axis. Although you may use Maple to check your answers to the questions below, determine your answers by geometric reasoning and explain your reasoning.
(a) Find the kernel and range of the composite transformation *P R*.
(b) Find the kernel and range of the composite transformation *R P*.

Student Workspace

Problem 5: Rotation about the diagonal of a cube (Challenge)

Consider the three special types of rotation matrices that rotate vectors about one of the coordinate axes:

```
> Rx := theta->rotatemat3d([1,0,0],theta);
> Ry := theta->rotatemat3d([0,1,0],theta);
> Rz := theta->rotatemat3d([0,0,1],theta);
>
```

The "unit cube" has one vertex at (0,0,0) and the opposite vertex at (1,1,1). Using only the three rotation matrices *Rx*, *Ry*, *Rz* above, construct the matrix that rotates the unit cube through the angle θ about the diagonal connecting (0,0,0) and (1,1,1) in the counter-clockwise direction when looking along the diagonal toward the origin. Test your transformation on the cube3d figure.

Hint: First rotate to line up the diagonal with one of the axes; then perform the desired rotation; and finally return the diagonal back to its original position.

Student Workspace

Pencil and Paper Practice Problems

Note: Maple is not required for any of the problems in this section. Do all of these by hand.

1. Find the 3 by 3 rotation matrix that rotates vectors 180 degrees about the line $y = x$ in the xy plane.

2. Find the kernel and range of the 3 by 3 projection matrix that projects vectors onto the plane $2x - y + 3z = 0$.

Linear Algebra Modules Project
Chapter 4, Module 3

Computer Graphics

Purpose of this module

The purpose of the module is to explore the uses of geometric matrix transformations in computer graphics. In particular, we introduce homogeneous coordinates as a means of combining translations with other geometric transformations on R^2 and R^3.

Prerequisites

Modules 1, "Geometry of Matrix Transformations of the Plane." Also, Module 2, "Geometry of Matrix Transformations of R^3," is required for Section 2.

Commands used in this module

```
> restart: with(plots): with(linalg): with(lamp):
>
```

Tutorial

Section 1. Homogeneous 2D Coordinates

The picture below shows a red triangle with vertices [4, 3], [5, 9], and [6, 3] that has been translated up and to the right by the vector $\begin{bmatrix} 3 \\ 2 \end{bmatrix}$. This is accomplished by adding the translation vector $\begin{bmatrix} 3 \\ 2 \end{bmatrix}$ to each of the points of the original figure.

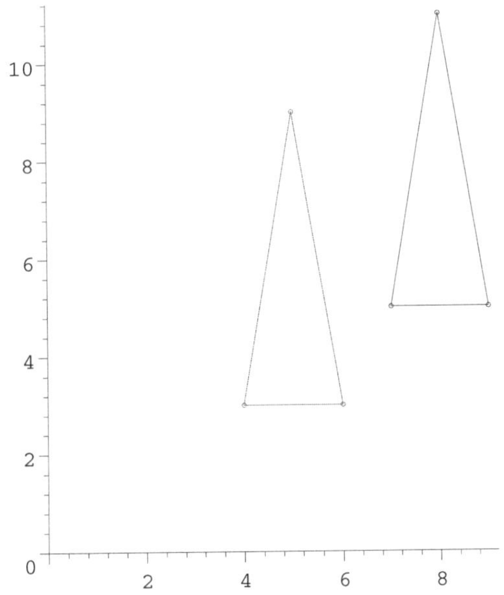

[>

Given our experience with geometric matrix transformations in Module 1, we naturally ask if we can accomplish such a translation by multiplying by a suitable 2 by 2 matrix. The answer turns out to be: No such matrix exists. Here is why. Imagine that you do have a matrix M that translates every point by the vector $\begin{bmatrix} 3 \\ 2 \end{bmatrix}$. Then you would get, for every vector $\begin{bmatrix} a \\ b \end{bmatrix}$ in R^2,

$$M\begin{bmatrix} a \\ b \end{bmatrix} = \begin{bmatrix} a+3 \\ b+2 \end{bmatrix} \text{ and hence, in particular, } M\begin{bmatrix} 0 \\ 0 \end{bmatrix} = \begin{bmatrix} 3 \\ 2 \end{bmatrix}.$$

On the other hand, if you multiply any matrix M by the zero vector, you always get the zero vector. So it is not possible for a 2 by 2 matrix M to accomplish this translation or indeed any translation.

However, in computer graphics animations, the most basic transformation is a translation. So we persist in asking: Is there some way in which we can use matrix multiplication both to translate figures and perform the other basic geometric transformations, such as dilations, rotations, etc.?

[>

Yes, there is a way -- but we need 3 by 3 matrices to do it! The trick is to express points in R^2 in what are called "homogeneous coordinates." A typical point $[x, y]$ becomes, in homogeneous coordinates, the point $[x, y, 1]$. While this looks like a point in R^3, we will treat it as a special coded form of the point $[x, y]$. So when we represent a figure by a matrix whose columns are points of the figure, we replace the former 2 by n matrix by a 3 by n matrix whose last row is all 1's.

For example, our red triangle above can be represented using homogeneous coordinates by the matrix "triangle":

```
[ > triangle := evalm(matrix([[4,5,6,4],[3,9,3,3],[1,1,1,1]]));
[ >
```

Geometric Transformations Using Homogeneous Coordinates

Under this new setup, the 2 by 2 geometric transformations from Module 1 now become 3 by 3 matrices. For example, here is the original version of the rotation matrix *R* followed by the new "homogeneous" version, which we have labeled *Rh*.

```
[ > R := rotatemat(theta);
[ > Rh := rotatemat(theta,homogeneous=on);
[ >
```

Note: To get the homogeneous version of the rotation matrix, we added the option "homogeneous=on" to the rotatemat(..) command. Also observe that the original rotation matrix makes up the 2 by 2 block in the upper-left corner of the homogeneous version. The rest of the matrix is filled out with zeros, except for a 1 in the lower-right corner.

Now compare how the two rotation matrices act on the vector $[x, y]$ and its homogeneous form $[x, y, 1]$.

```
[ > p := colvector([x,y]);
    ph := colvector([x,y,1]);
[ > evalm(R&*p);
[ > evalm(Rh&*ph);
```

If we drop the 1 in the homogeneous form, we get the same image point by both methods.

=====================

Example 1A: Use homogeneous coordinates to rotate the "triangle" figure by an angle of $\pi/4$ radians.

First we find the homogeneous version of the appropriate rotation matrix:

```
[ > M := rotatemat(Pi/4,homogeneous=on);
```

Next we apply the transform command. Notice that this command also contains the option "homogeneous=on".

```
[ > transform(M,triangle,homogeneous=on);
[ >
```

=====================

Note: Adding "homogeneous=on" to each command can become quite tedious. Alternatively, you can turn the homogeneous mode on for all future commands by executing the next line.

```
[ > homogeneous := on;
[ >
```

You can also override this global setting in any single command by including the option "homogeneous=off". To turn the homogeneous mode off entirely, use the command

> homogeneous := off;

All the other geometric transformations from Module 1 have homogeneous versions. Here are the reflection and projection matrices in homogeneous coordinates:

```
[ > reflectmat(theta);
[ > projectmat(theta);
```

====================

Example 1B: We use the homogeneous version of the reflection matrix to reflect the triangle figure about the y axis:

```
[ > M := reflectmat(Pi/2);
[ > transform(M,triangle);
[ >
```

====================

Translation Using Homogeneous Coordinates

Of course the reason for switching to homogeneous coordinates is to enable us to treat translation as a matrix product. So here is the matrix that translates points by the vector $\begin{bmatrix} h \\ k \end{bmatrix}$:

```
[ > T := translatemat([h,k]);
```

And here is proof that the translation matrix does indeed translate each point $[x, y]$ (in homogeneous coordinates) to $[x+h, y+k]$ (in homogeneous coordinates):

```
[ > P := colvector([x,y,1]);
[ > evalm(T&*P);
```

Now let's translate the red triangle in the introduction by the vector $\begin{bmatrix} 3 \\ 2 \end{bmatrix}$:

```
[ > A := translatemat([3,2]);
[ > transform(A,triangle);
[ >
```

The translation matrix also provides us with much greater flexibility in applying the other geometric transformations. For example, all the rotations in Module 1 have the origin as the center of the rotation, but now we can rotate an object about any other center point.

====================

Example 1C: Rotate the triangle figure counterclockwise 45 degrees about its top vertex, [5,9].

Solution: We break the problem into the following three steps.

Step 1. Translate the triangle so the vertex [5,9] is at the origin.
Step 2. Rotate the translated figure 45 degrees.
Step 3. Translate the figure back.

```
[ >
```

Each of these steps is accomplished (see below) by applying a matrix in homogeneous coordinates. The product of the three matrices gives us a single matrix that accomplishes the desired rotation in a single step.

Step 1. Translate the triangle so the vertex [5,9] is moved to the origin. We therefore translate the triangle by the vector $\begin{bmatrix} -5 \\ -9 \end{bmatrix}$.

```
[ > T := translatemat([-5,-9]);
[ > transform(T,triangle,view=[-10..10,-10..10]);
[ >
```

Step 2. Rotate the translated figure 45 degrees. Apply this rotation matrix *R* to the translated triangle, which we call "triangle1".

```
[ > R := rotatemat(Pi/4);
[ > triangle1 := evalm(T&*triangle);
[ > transform(R,triangle1,view=[-10..10,-10..10]);
[ >
```

Step 3. Translate the figure so the origin is moved back to [5,9]. Apply this translation matrix *U* to the rotated triangle, which we call "triangle2". (Note that the translation matrices *T* and *U* are inverses of one another.)

```
[ > U := translatemat([5,9]);
[ > triangle2 := evalm(R&*triangle1);
[ > transform(U,triangle2,view=[-10..10,-10..10]);
[ >
```

Now we multiply all three matrices together and thereby construct a single matrix that performs the rotation about the vertex [5, 9] with a single multiplication.

```
[ > M := evalm(U&*R&*T);
[ > transform(M,triangle,view=[-10..10,-10..10]);
[ >
```

====================

Making Movies

We can also apply the product matrix *M* repeatedly to create a simple animation:

```
[ > movie(M,triangle,20);
[ >
```

Execute the next line to see the individual frames used in the animation.

```
[ > movie(M,triangle,20,frames=on);
[ >
```

Exercise 1.1: Modify the commands in Example 1C to do each of the following:

(a) Rotate the triangle figure about its top vertex by a smaller angle, say $\pi/8$. Run movie(..) to check your matrix. (The commands that created the matrix *M* in Example 1C are repeated in the Student Workspace. Start by modifying those commands.)

(b) Rotate about one of the other vertices by the same angle as in (a) but rotate clockwise. Remember to adjust both translation matrices.

(c) The "lamph" figure is shown below. Find a single matrix *N* that rotates this figure by 90 degrees about the center point [7, 7.5]. Test your matrix by using the transform(..) command.

```
[ > drawmatrix(lamph,figcolor=blue);
[
```

```
[ >
```

Student Workspace

Answer 1.1

Section 2. Homogeneous 3D Coordinates

We define the homogeneous coordinates for a vector $[x, y, z]$ in R^3 to be $[x, y, z, 1]$. The geometric transformations of R^3 are then represented by 4 by 4 matrices. For example, here are the scaling and reflection matrices from Module 2, side by side with their homogeneous versions. Notice that in each case the original transformation makes up the 3 by 3 block in the upper-left corner of the matrix. The rest of the matrix is filled out with zeros, except for a 1 in the lower-right corner.

$$\begin{bmatrix} 2 & 0 & 0 \\ 0 & 3 & 0 \\ 0 & 0 & 5 \end{bmatrix} \quad \begin{bmatrix} 2 & 0 & 0 & 0 \\ 0 & 3 & 0 & 0 \\ 0 & 0 & 5 & 0 \\ 0 & 0 & 0 & 1 \end{bmatrix} \qquad \begin{bmatrix} -1 & 0 & 0 \\ 0 & 1 & 0 \\ 0 & 0 & 1 \end{bmatrix} \quad \begin{bmatrix} -1 & 0 & 0 & 0 \\ 0 & 1 & 0 & 0 \\ 0 & 0 & 1 & 0 \\ 0 & 0 & 0 & 1 \end{bmatrix}$$

Since we will be working with homogeneous coordinates throughout this section, we set homogeneous := on:

```
[ > homogeneous := on;
```

Throughout this section we will apply matrices to the figure "hotel3dh," which is the hotel3d figure in Module 2 expressed in homogeneous coordinates. Move it around to become familiar with its shape and location:

```
[ > drawmatrix3d(hotel3dh);
[ >
```

Note: The names of all figures that we use in the remainder of this module will end in an "h" to signify that they are expressed in homogeneous coordinates. Some of the 2D figures that you used in Module 1 have been redefined as figures lying in the *xy* plane of R^3. For example, here is the "bug" sitting in the *xy* plane of R^3:

```
[ > drawmatrix3d(bug3dh);
```

Other figures are "house3dh", "lamp3dh", and a new figure, "chevron3dh", that lies in the *xz* plane:

```
[ > drawmatrix3d(chevron3dh);
[ >
```

====================

Example 2A: The 3D translation matrix is similar to the one we used in Section 1. We construct the matrix corresponding to translation by the vector $\begin{bmatrix} a \\ b \\ c \end{bmatrix}$ by using translatemat3d([a,b,c]) . For example:

```
[ > T := translatemat3d([10,-6,8]);
```

If we multiply the homogeneous coordinate vector $[x, y, z, 1]$ on the left by this matrix, we obtain the vector translated by $[10, -6, 8]$:

```
[ > P := colvector([x,y,z,1]);
[ > evalm(T&*P);
```

Now we apply T to hotel3dh by using the transform3d(..) command. Move the plot around until you can see the translation from a good point of view. Try looking at it straight down each of the coordinate axes.

```
[ > transform3d(T,hotel3dh);
[ >
```

====================

Rotation Transformations in R^3 Using Homogeneous Coordinates

As in Module 2, we use the command rotatemat3d(..) to produce rotation matrices for R^3. For example, to produce the nonhomogeneous and homogeneous version of the matrix that rotates vectors about the z axis, we use:

```
[ > rotatemat3d([0,0,1],theta,homogeneous=off);
[ > rotatemat3d([0,0,1],theta,homogeneous=on);
[ >
```

====================

Example 2B: Let's rotate the chevron3dh figure by 30 degrees about the z axis. (Since the homogeneous mode is still "on" globally we do not need to include it in each command.)

```
[ > Rz := rotatemat3d([0,0,1],Pi/6);
[ > transform3d(Rz,chevron3dh);
[ >
```

Next, let's use the movie3d(..) command to see what happens when this rotation is applied 12 times:

```
[ > movie3d(Rz,chevron3dh,12);
[ >
```

====================

Explore ~~ Here is an animation based on a composition of two matrices. Can you predict what the animation will look like before running it? Try modifying the commands to reverse the direction of the rotation or the translation (or both). If you are working at a fast computer, you might try increasing the number of frames from 12 to 24.

```
[ > T := translatemat3d([0,0,1]);
[ > Rz := rotatemat3d([0,0,1],Pi/6);
[ > M := evalm(Rz&*T);
[ > movie3d(M,chevron3dh,12);
[ >
```

====================

Example 2C: We rotate the hotel3dh figure about the y axis, first by a single rotation, then by the same rotation applied over and over:

```
[ > Ry := rotatemat3d([0,1,0],Pi/12);
[ > transform3d(Ry,hotel3dh);
[
```

```
> movie3d(Ry,hotel3dh,12);
>
```

===================

Exercise 2.1: Find a 4 by 4 matrix that reflects vectors across the *yz* plane using homogeneous coordinates. Test your matrix on the hotel3dh figure.

Student Workspace

Answer 2.1

Exercise 2.2: Find a 4 by 4 matrix that projects vectors onto the *xz* plane. Test your matrix on the hotel3d figure.

Student Workspace

Answer 2.2

Section 3. Moving an Object Along a Curve (Optional)

If we want to create a realistic animation of a car moving along a curved road, we face two problems:

- How can we keep the car on the road as the road curves?
- How can we keep the car pointing in the direction of motion?

We address these problems one at a time in the following two subsections. First, let's again be certain that we are in the homogeneous mode.

```
> homogeneous := on;
>
```

Staying on the Curve

Here is the parametrized curve we will use:

```
> x := t -> cos(t);
  y := t -> sin(2*t);
```

And here is the definition of the car in homogeneous coordinates:

```
> car :=
  matrix([[.15,-.15,-.15,.15],[0,.075,-.075,0],[1,1,1,1]]);
```

We plot both the curve and the car in a single picture. Note that the car is pointing in the direction of the positive *x* axis.

```
> pict1 := drawmatrix(car):
  pict2 := plot([x(t),y(t),t=0..2*Pi],color=black):
  display([pict1,pict2]);
>
```

To create the illusion of motion along the curve, we need to place the car at numerous points along the curve. We do this by translating the car from the origin to points on the curve. For example, let's translate the car to the point corresponding to time $t = 2$:

```
> T := translatemat([x(2),y(2)]);
```

```
> pict1 := drawmatrix(evalm(T&*car)):
  pict2 := plot([x(t),y(t),t=0..2*Pi],color=black):
  display([pict1,pict2]);
>
```

Next, we write a loop to create and save a similar picture for 16 values of t, ranging from $t = 0$ to $t = 3.0$ in increments of .2 . We call the pictures pic_1, pic_2, etc., so we can display them after we have created them.

Maple Note: When writing a "do loop", you begin with a "for" statement that tells how many times to execute the loop. The rest of the loop consists of a sequence of Maple commands preceded by "do" and followed by "od:" All the lines of code should follow a single Maple prompt (>).

```
> for k from 0 to 15
  do
  tvalue := .2*k:
  T := translatemat([x(tvalue),y(tvalue)]);
  part1 := drawmatrix(evalm(T&*car)):
  part2 := plot([x(t),y(t),t=0..2*Pi],color=black):
  pic[k+1] := display([part1,part2]):
  od:
>
```

Exercise 3.1: Explain in words what each of the five lines in the loop does.

Student Workspace

Answer 3.1

```
>
```

Now we can see the pictures all at once by using the display(..) command:

```
> display([seq(pic[k],k=1..16)]);
>
```

Or, by adding the instruction "insequence=true" to the previous display(..) command, we can create an animation:

```
> display([seq(pic[k],k=1..16)],insequence=true);
>
```

Pointing in the Direction of Motion

We are halfway there! Now to keep the car pointing in the direction of motion, we should rotate the car at each step of the animation. If we can find the appropriate rotation matrix R for each value of t, we will have solved our last remaining problem.

Recall that the velocity (or tangent) vector always points in the direction of motion. We calculate the velocity vector by taking the derivative of each of the component functions with respect to t:

```
> vel := colvector([diff(x(t),t),diff(y(t),t)]);
```

Next we define the unit vector T in the direction of vel, namely $vel/|vel|$:

```
[ > T := evalm(vel/mag(vel));
```

Recall that the columns of a 2 by 2 matrix transformation are the images of the standard unit vectors e_1 and e_2. Since the car points in the direction of e_1 and we want it to point in the direction of the unit vector T, the image of e_1 is T. Furthermore, if we were to rotate e_1 by 90 degrees counter-clockwise, we would get e_2; so rotating T by 90 degrees counter-clockwise, we get the image of e_2. Let's call this image N:

```
[ > N := evalm(rotatemat(Pi/2,homogeneous=off)&*T);
```

The rotation matrix we seek has columns T and N:

```
[ > R := augment(T,N);
```

We must also construct the corresponding homogeneous version of this rotation matrix:

```
[ > Rh := copyinto(R,identmat(3),1,1);
```

Since the rotation matrix is a function of t, let's express it as such:

```
[ > Rot := tval->map(evalf,subs(t=tval,evalm(Rh)));
```

For example, here is the rotation matrix when $t = 2$:

```
[ > Rot(2);
[ >
```

We are ready to test our results. We repeat the loop that we used earlier with one modification. Can you see where we have made the change?

```
[ > for k from 0 to 15
    do
    tvalue := .2*k:
    U := translatemat([x(tvalue),y(tvalue)]);
    part1 := drawmatrix(evalm(U&*Rot(tvalue)&*car)):
    part2 := plot([x(t),y(t),t=0..2*Pi],color=black):
    pic[k+1] := display([part1,part2]):
    od:
[ >
```

Now try out the animation:

```
[ > display([seq(pic[k],k=1..16)],insequence=true);
[ >
```

You can see the individual elements of the picture by executing the next line.

```
[ > display([seq(pic[k],k=1..16)]);
[ >
```

Problem 5 asks you to apply the ideas in this section to motion along a curve in R^3. However, for that problem, you will also need to know how to use the unit normal vector to find the rotation matrix. That approach is addressed in the following appendix.

Appendix. An Alternative Method: Using the Unit Normal Vector

We introduce in this section an alternative method for finding the rotation matrix *Rot* we found above. Recall from calculus that the unit normal vector N for a curve is perpendicular to the unit tangent vector T. Furthermore, N is found by computing the derivative of T and then converting the derivative to a unit vector. This is quite tedious to do by hand, but routine for Maple. In the region below, we calculate the derivative of T and call it dT. Then we define the unit normal vector $N = dT/|dT|$. We have suppressed the output of these expressions since they are quite messy. Change the colons to semicolons if you would like to see the output.

```
> dT := map(diff,T,t):
> N := evalm(dT/mag(dT)):
>
```

Finally, we build the rotation matrix *Rot* from T and N in much the same way we did earlier:

```
> R := augment(T,N):
> Rh := copyinto(R,identmat(3),1,1):
> Rot := tval->map(evalf,subs(t=tval,evalm(Rh)));
>
```

Problems

Problem 1: What matrix produces the movie?

The 2D animation below was created using the command

movie(M,triangle,18)

for some matrix M. The triangle dimensions are decreasing by 10% in each frame. Find the matrix M. Hint: M is the product of four matrices. In your answer, list the four matrices that are used to construct M, and indicate the order in which they are multiplied. Check your answer by constructing M and then entering and executing the above movie(..) command.

```
> triangle := evalm(matrix([[4,5,6,4],[3,9,3,3],[1,1,1,1]]));
>
```

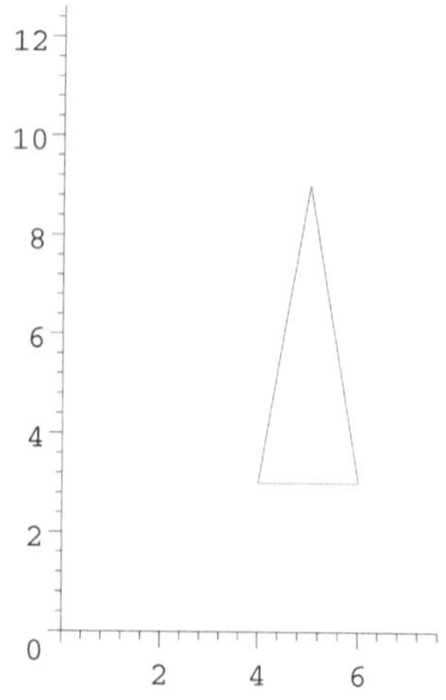


```
>
```

Student Workspace

Problem 2: Translation and rotation

(a) Enter the 3 by 4 matrix *tri* whose columns are the homogeneous coordinates for the triangle with vertices $[-1, -1]$, $[1, -1]$, $[0, 1]$. Use the command drawmatrix(tri,homogeneous=on) to draw the triangle.
(b) Use translatemat(..) and transform(..) to show the triangle *tri* translated to the right 10 units.
(c) (By hand) Find the vertices of the translated triangle in (c) by multiplying the homogeneous coordinates in (a) by the translation matrix in (b).
[>
(d) Make a 36-frame animation (using movie(..)) of the triangle *tri* in which each frame is the result of applying a rotation of $\pi/18$ radians followed by a translation by the vector $\begin{bmatrix} .4 \\ .1 \end{bmatrix}$.
(e) (Challenge) Explain why, in the part (d) animation, the triangle *tri* returns to its original position; that is, explain why the repeated translations do not translate the triangle.

Student Workspace

Problem 3: Twister -- the movie

The animation below was created using the command movie3d(M,hotel3dh,16) for some matrix *M*. Find *M*. Hint: You may want to view the animation from several points of view. It may also help to advance it frame by frame using the button marked ->| .
[>

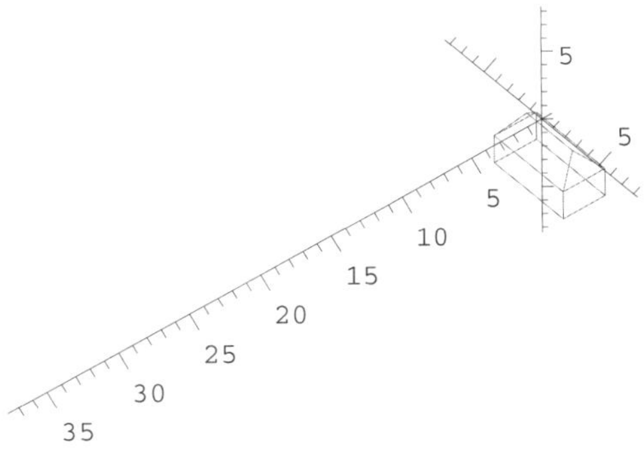

[>

Student Workspace

Problem 4: Programming a 3D animation (Optional)

Note: Study the discussion of programming in Section 3 before attempting this problem.

Execute the all the commands below to generate an animation. After studying the animation, answer the questions that follow.

```
> homogeneous := on:
> p := 'p';
> L := diag(1,0,1,1);
> M := diag(0,1,1,1);
> N := translatemat3d([-3.5,0,-2]);
> for j from 0 to 12
  do
  R := rotatemat3d([0,1,0],j*Pi/6):
  W := evalm(N^(-1)&*R&*N&*hotel3dh):
  part1 := drawmatrix3d(W,linecolor=red):
  part2 := drawmatrix3d(evalm(L&*W),linecolor=blue):
  part3 := drawmatrix3d(evalm(M&*W),linecolor=blue):
  p[j+1] := display([part1,part2,part3]):
  od:
> display([seq(p[k],k=1..13)],insequence=true);
>
```

(a) Describe in words what is happening in the animation.
(b) Explain the purpose for each of the commands above.
(c) Modify the code to create a similar animation based on a rotation about a line that is parallel to the *x* axis (instead of the *y* axis) and that cuts the hotel3d figure through a central point.

Student Workspace

Problem 5: Flying along a curve in 3-Space (Optional)

In this problem you will extend the methods in Section 3 to the corresponding problem of moving a figure along a curve in 3-space. You are given a space curve and a figure called jet3dh. The figure is oriented so that its wings are in the *xy* plane and its nose is in the direction of e_1. Your task is to create an animation that flies jet3dh along the curve, subject to the following constraints:

At each point of the curve, jet3dh should:
(a) point in the direction of the tangent vector, and
(b) have its wings lie on the "osculating plane" (the plane containing the unit tangent vector T and unit normal vector N).

Hints: Start with a copy of the commands in Section 3 and make whatever modifications are required. To draw a space curve, use the spacecurve(..) command. To show jet3dh without its corner points, use the option points=off in the drawmatrix3d(..) command. Calculate N as in the appendix to Section 3. It will help to calculate a third unit vector $B = T \times N$ (the cross product of T and N).

Here is the curve that the jet3dh will fly along:

```
> x := t -> cos(t)/2;
  y := t -> sin(t);
  z := t -> sin(2*t)/2;
```

We plot both the curve and jet3dh in a single picture.

```
> part1 :=
  drawmatrix3d(jet3dh,points=off,linecolor=blue,axes=framed):
  part2 := spacecurve([x(t),y(t),z(t),t=0..2*Pi],color=red):
  display3d([part1,part2]);
> 
```

Problem 6: The Art of linear algebra (Optional)

Was there a picture or animation in this module that you found particularly appealing? Did it make you want to try to create something of your own ? This problem is an invitation for you to be creative with the mathematical tools that have been presented in this module. The only requirement is that you annotate your work so that someone else (such as another student) can understand what you did.

Student Workspace

```
> 
```

Pencil and Paper Practice Questions

Note: Maple is not required for any of the problems in this section. Do all of these by hand.

1. (a) If $\begin{bmatrix} a & b \\ c & d \end{bmatrix}$ is a matrix transformation, write down the corresponding homogeneous version.

(b) Show that the effect of the nonhomogeneous matrix in (a) on $[x, y]$ and the effect of the corresponding homogeneous matrix on $[x, y, 1]$ are the same (if we drop the extra 1 in the homogeneous case).

2. Suppose we want to reflect a figure in R^2 across the vertical line $x = 2$. Find three matrices of the sort studied in this module whose composite will perform this reflection. Do not compute their product, but do indicate the order in which they are to be multiplied.

Linear Algebra Modules Project
Chapter 5, Module 1

Subspaces of R^n

Special instructions

The Tutorial for this module is designed to be done by hand without the use of Maple. Likewise, most of the Problems can and should be done without the use of Maple. However, feel free to use Maple to try examples and to check your answers. The answers to the Exercises are collected together at the end of the Tutorial.

Purpose of this module

The purpose of this module is to introduce the concept of a subspace of R^n and to study the principal examples of subspaces.

Prerequisites

```
> restart: with(linalg): with(lamp):
>
```

Tutorial

Section 1: Definition of Subspace

To begin our discussion of subspaces we consider the simple example of a plane passing through the origin in R^3.

```
>
```

================

Example 1A: Consider the plane through the origin defined by the homogeneous equation $x - 3y + 2z = 0$. For convenience we will use the letter S to denote the set consisting of all vectors in this plane. In other words, S is the solution set for this homogeneous equation.

We decompose the solution set S, and thus find two direction vectors that span the plane:

$$\begin{bmatrix} x \\ y \\ z \end{bmatrix} = \begin{bmatrix} 3y - 2z \\ y \\ z \end{bmatrix} = y \begin{bmatrix} 3 \\ 1 \\ 0 \end{bmatrix} + z \begin{bmatrix} -2 \\ 0 \\ 1 \end{bmatrix} = y\, v_1 + z\, v_2$$

$$\text{where } v_1 = \begin{bmatrix} 3 \\ 1 \\ 0 \end{bmatrix} \text{ and } v_2 = \begin{bmatrix} -2 \\ 0 \\ 1 \end{bmatrix}$$

The vectors v_1, v_2 give us an alternative way of describing S, namely S = Span{ v_1, v_2 }. This description of the plane has the advantage that we can use v_1 and v_2 to construct all of the vectors in the plane. In contrast, the equation $x - 3\,y + 2\,z = 0$ describes the plane S only implicitly; that is, this equation allows us to test whether any vector is in the plane, but it doesn't give us a direct way to construct these vectors.

================

[>

We now note two simple, but important, algebraic properties that hold for vectors in the set S:

- (1) If you choose any vector u in S and multiply it by any real number k, the resulting vector $k\,u$ will also be in S.
- (2) If you choose any two vectors u and v in S and add them, the resulting vector $u + v$ will also be in S.

Try visualizing these properties using the geometry of vectors in 3-space.

[>

Proving these properties is straightforward using vector algebra :

Proof of (1): Let u be any vector in S and k be any scalar. Since S = Span{ v_1, v_2 }, we can express u as a linear combination of the vectors v_1 and v_2: $u = a\,v_1 + b\,v_2$ (where a and b are real numbers). Now multiply this equation by k; we get

$$k\,u = k\,(a\,v_1 + b\,v_2) = (k\,a)\,v_1 + (k\,b)\,v_2$$

So $k\,u$ is also a linear combination of the vectors v_1 and v_2, and hence $k\,u$ is contained in S.

Proof of (2): Let u and v be any vectors in S. Each can be expressed as a linear combination of v_1 and v_2: $u = a\,v_1 + b\,v_2$ and $v = c\,v_1 + d\,v_2$. Now add the two equations; we get

$$u + v = (a\,v_1 + b\,v_2) + (c\,v_1 + d\,v_2) = (a + c)\,v_1 + (b + d)\,v_2$$

So $u + v$ is also a linear combination of v_1 and v_2, and hence $u + v$ is contained in S.

[>

The two properties above are the defining characteristics of a "subspace" of R^n.

> ***Definition:*** Let S be a set of vectors in R^n. S is a ***subspace*** of R^n if:
> (1) $k\,u$ is in S whenever u is a vector in S and k is a scalar; and
> (2) $u + v$ is in S whenever u and v are vectors in S; and
> (3) The zero vector of R^n is contained in S.

[>

Conditions (1) and (2) are often referred to as the "closure" properties for a subspace. We say that that a subspace is "closed" under the operations of scalar multiplication and vector addition in the sense that these operations always produce vectors that are still within the subspace. Condition (3) is included simply to guarantee that a subspace contains at least one vector; an empty subspace would be of no use.

Our first example of a subspace, the plane through the origin in Example 1A, turns out to be a very good model for many of the subspaces that you will encounter later in this chapter. The next two theorems describe the two ways in which subspaces typically arise.

[>

Theorem 1: If $v_1, v_2, ..., v_p$ are vectors in R^n, then $\text{Span}\{v_1, v_2, ..., v_p\}$ is a subspace of R^n.

Exercise 1.1: The proof of Theorem 1 is a natural generalization of the approach we used in proving the closure properties (1) and (2) for $\text{Span}\{v_1, v_2\}$ in Example 1A. Using the proof from Example 1A as a model, write down the proof for the case where $p = 3$; i.e., show that if v_1, v_2, v_3 are any three vectors in R^n, then $\text{Span}\{ v_1, v_2, v_3 \}$ is a subspace of R^n.

[>

Student Workspace

Answer 1.1

Now let's generalize the first way we described the subspace S in Example 1A, namely by a homogeneous linear equation. The generalization is the solution set of any homogenous linear system:

Theorem 2: Let A be any m by n matrix. The set of solutions of the homogeneous equation $A\,x = 0$ is a subspace of R^n.

At first glance, Theorems 1 and 2 may seem quite unrelated. However, there is a close connection between them, as Exercise 1.2 illustrates.

Exercise 1.2: Let A be the 3 by 4 matrix shown below.

```
[ > A := matrix([[1,3,0,4],[1,2,2,-4],[2,6,0,8]]);
[ >
```

$$A = \begin{bmatrix} 1 & 3 & 0 & 4 \\ 1 & 2 & 2 & -4 \\ 2 & 6 & 0 & 8 \end{bmatrix}$$

(a) Solve the homogeneous equation $A\,x = 0$, and decompose the solution set S into the span of a set of vectors.

(b) Use the result from (a) to show that S is a subspace of R^4.

[>

Student Workspace

Answer 1.2

It is easy to see how the procedure used in answering Exercise 1.2 could be applied to any homogeneous equation $A\,x = 0$. But how could we prove Theorem 2 directly from the definition of subspace? In Problem 1 you will use the rules of matrix algebra to write a short proof of Theorem 2 without solving the homogeneous system or using Theorem 1.

Summary: The principal ways in which subspaces of R^n arise are as the span of a set of vectors and as the solution set of a homogeneous linear system.

(In fact, any subspace of R^n can be expressed in either of these two ways.)

[>

================

Example 1B: Suppose S is a subspace of R^n, and suppose we choose four vectors from this subspace at random. Call these vectors w_1, w_2, w_3, w_4. Now we form a new vector p by taking a linear combination of these four vectors. For example, let $p = 2\,w_1 + 5\,w_2 + 4\,w_3 + 7\,w_4$. Explain why p must also be a vector in the subspace S.

Solution: Each of the four terms in the linear combination is a vector in S since each term is a scalar mulitple of a vector in S. So let's rewrite p as the sum of these four vectors, $p = y_1 + y_2 + y_3 + y_4$, where $y_1 = 2\,w_1$, $y_2 = 5\,w_2$, etc. We need to show that the sum of these four vectors in S is a vector in S. The sum of the first two vectors, $y_1 + y_2$, and the sum of the last two vectors, $y_3 + y_4$, are each vectors in S by the closure property for addition. So we can rewrite p as $p = z_1 + z_2$, where $z_1 = y_1 + y_2$ and $z_2 = y_3 + y_4$ are each vectors in S. Therefore p is a vector in S, since it is a sum of two vectors in S.

================

[>

Example 1B illustrates a general fact about subspaces that we summarize as follows:

- If S is a subspace of R^n and $\{v_1, v_2, ..., v_p\}$ is any set of vectors in S, then every linear combination of these vectors is in S.

[>

Section 2: Geometry of Subspaces

What subsets of R^2 are subspaces of R^2? What subsets of R^3 are subspaces of R^3? We will see that the only subsets that are subspaces are the "linear" subsets through the origin, such as lines and planes through the origin. This leaves a great many subsets that are not subspaces. For example, subsets of R^2 that are not subspaces include curves that are not lines, lines that do not go through the origin, or any bounded set (such as a circle, polygon, or line segment) except for the special example $\{0\}$.

[>

================

Example 2A: Find all possible subspaces of R^2.

Solution: We will build up all possible subspaces systematically from the smallest to the largest.
Case 1: The set $\{0\}$ containing just the zero vector of R^2 is a subspace of R^2: (1) Any scalar times the zero vector is the zero vector; (2) the zero vector plus the zero vector is the zero vector.

Case 2: If a subspace S contains something more than just the zero vector, we can select a nonzero vector u in S. Then, by property (1) for subspaces, S must contain all vectors $a\,u$, where a is any scalar. That is, S must contain all vectors in Span$\{u\}$. If S = Span$\{u\}$, then S is a line through the origin, which, by Theorem 1, is a subspace of R^2.

[>

Case 3: If a subspace S contains something more than just a line through the origin, i.e., more than Span$\{u\}$, we can select a vector v that is in S but not in Span$\{u\}$. Since S is a subspace, S must contain Span$\{u, v\}$. Since v is not in Span$\{u\}$, the vectors u and v are not parallel; therefore Span $\{u, v\}$ is all of R^2. So S = R^2.

Conclusion: The subspaces of R^2 are: The set $\{0\}$ containing just the zero vector; the lines through the origin (that is, each such line is a subspace); and all of R^2.

================

In Problem 2, we will ask you to carry out a similar analysis for subspaces of R^3.

[>

Exercise 2.1: For each of the following subsets of R^2, show that the subset is not a subspace. Use the definition of subspace rather than the result of Example 2A (so you will get practice using the definition). To show that a subset is not a subspace, give a counter-example. Also draw the subset (by hand, if you like), and draw in the vectors that provide your counter-example.

(a) The set of all points on the line $x + y = 1$.

(b) The set of all points on the half-line $y = x$, where $0 \le x$.

(c) The set of all points on the graph of the equation $y = x^2$.

[>

Student Workspace

Answer 2.1

[>

Answers to Exercises

Answer 1.1:

Let v_1, v_2, v_3 be any three vectors in R^n. We show that S = Span{ v_1, v_2, v_3 } is a subspace of R^n by showing that the three conditions of the definition are satisfied.

[>

Proof of (1): Let u be any vector in S and k be any scalar. Since S = Span{ v_1, v_2, v_3 }, we can express u as a linear combination of the vectors v_1, v_2, v_3: $u = a\,v_1 + b\,v_2 + c\,v_3$ (where a, b, c are real numbers). Now multiply this equation by k; we get:

$$k\,u = k\,(a\,v_1 + b\,v_2 + c\,v_3) = (k\,a)\,v_1 + (k\,b)\,v_2 + (k\,c)\,v_3$$

So $k\,u$ is also a linear combination of the vectors v_1, v_2, v_3, and hence $k\,u$ is contained in *S*.

[>

Proof of (2): Let u and v be any vectors in S. Each can be expressed as a linear combination of v_1, v_2, v_3: $u = a\,v_1 + b\,v_2 + c\,v_3$ and $v = d\,v_1 + e\,v_2 + f\,v_3$. Now add these two equations; we get:

$$u + v = (a\,v_1 + b\,v_2 + c\,v_3) + (d\,v_1 + e\,v_2 + f\,v_3) = (a + d)\,v_1 + (b + e)\,v_2 + (c + f)\,v_3$$

So $u + v$ is also a linear combination of v_1, v_2, v_3, and therefore $u + v$ is a vector in S.

Proof of (3): The zero vector is included in the span of any set of vectors, as we see by choosing all the weights to be zero: $0\,v_1 + 0\,v_2 + 0\,v_3 = 0$.

[>

Answer 1.2:

Although we could solve $A\,x = 0$ entirely by hand, here is a solution aided by Maple:

```
[ > A := matrix([[1,3,0,4],[1,2,2,-4],[2,6,0,8]]);
[ > zero := colvector([0,0,0]);
[ > x := matsolve(A,zero);
[ >
```

(a) The solution produced by the matsolve(..) command is

$$x = \begin{bmatrix} -6\,t_2 + 20\,t_1 \\ 2\,t_2 - 8\,t_1 \\ t_2 \\ t_1 \end{bmatrix}$$

Decomposing this solution set, we find that every solution can be expressed as a linear combination of two vectors, v_1 and v_2, in R^4:

$$x = t_1 \begin{bmatrix} 20 \\ -8 \\ 0 \\ 1 \end{bmatrix} + t_2 \begin{bmatrix} 6 \\ 2 \\ 1 \\ 0 \end{bmatrix} = t_1\, v_1 + t_2\, v_2$$

(b) Since S = Span$\{v_1, v_2\}$, where v_1 and v_2 are the two vectors above, S is a subspace of R^4 by Theorem 1.

[>

Answer 2.1:

(a) The vector $\begin{bmatrix} 0 \\ 0 \end{bmatrix}$ is not in the subset, which violates condition (3) of the definition. (Conditions (1) and (2) are also violated.)

(b) The vector $\begin{bmatrix} 1 \\ 1 \end{bmatrix}$ is in the subset, but $-\begin{bmatrix} 1 \\ 1 \end{bmatrix}$ is not in the subset, which violates condition (1) of the definition. (Neither condition (1) nor (2) is violated.)

(c) The vectors $\begin{bmatrix} 1 \\ 1 \end{bmatrix}$ and $\begin{bmatrix} -1 \\ 1 \end{bmatrix}$ are in the subset, but their sum $\begin{bmatrix} 0 \\ 2 \end{bmatrix}$ is not in the subset, which violates condition (2) of the definition. (Condition (1) is also violated but not (3).)

[>

Problems

Problem 1: Solution set of $A\,x = 0$ is a subspace

Prove Theorem 2 directly from the definition of subspace. That is, let A be an m by n matrix, and let S be the set of all vectors x in R^n such that $A\,x = 0$. Use the rules of matrix algebra to show that S satisfies conditions (1), (2), and (3).

Note: To help you see the type of proof that is wanted here, we provide the first few steps of the proof of condition (2):
Let u and v be any vectors in S. Then, $A\,u = 0$ and $A\,v = 0$. Now use the rules of matrix algebra to show that $u + v$ is in S. (Use a similar method for the proof of condition (1).)

Student Workspace

Problem 2: Subspaces of R^3

What are all the subspaces of R^3? Model your method of analysis on the method in Example 1A.

Student Workspace

Problem 3: Subsets that are not subspaces

For each of the following subsets of R^n, show that the subset is not a subspace by giving a counter-example to one of the parts of the definition of subspace. A drawing might help your explanation.

(a) The subset of R^3 consisting of all the solutions to the equation $2x - 3y + 4z = 5$.

(b) The subset of R^2 consisting of all the rational points on the line $y = x$; that is, x (and hence y) is any ratio of integers, m/n ($n \neq 0$).

(c) The subset of R^2 consisting of all the points on one or both of the lines $y = x$ and $y = -x$ (i.e., the "union" of these two lines).

(d) The subset of R^5 consisting of all the solutions of the nonhomogeneous equation $A x = b$, where A is a given 4 by 5 matrix and b is a given nonzero vector in R^4.

Student Workspace

Problem 4: Two ways of representing a subspace

Let $S_1 = \text{Span}\{u, v\}$, where u and v are given below, and let S_2 be the solution set of the homogeneous equation $A x = 0$, where A is given below.

(a) Determine whether the vectors p and q below are in S_1 or not; also determine whether they are in S_2 or not.

(b) Show that S_1 and S_2 are the same subspace; that is, show that every vector in S_1 is also in S_2, and every vector in S_2 is also in S_1.

```
> u := colvector([2,-1,0,2]);
  v := colvector([0,1,-4,2]);
> A := matrix([[0,2,1,1],[2,2,0,-1]]);
> p := colvector([1,-1,2,0]);
  q := colvector([1,0,2,0]);
>
```

$$u = \begin{bmatrix} 2 \\ -1 \\ 0 \\ 2 \end{bmatrix} \quad v = \begin{bmatrix} 0 \\ 1 \\ -4 \\ 2 \end{bmatrix} \quad p = \begin{bmatrix} 1 \\ -1 \\ 2 \\ 0 \end{bmatrix} \quad q = \begin{bmatrix} 1 \\ 0 \\ 2 \\ 0 \end{bmatrix}$$

$$A = \begin{bmatrix} 0 & 2 & 1 & 1 \\ 2 & 2 & 0 & -1 \end{bmatrix}$$

Student Workspace

Problem 5: Vectors $A\ x$ form a subspace

(a) Let A be the 4 by 3 matrix defined below. Let S denote the subset of R^4 consisting of all the vectors $A\ x$, where x ranges over all the vectors in R^3. Explain why S is a subspace of R^4. Hint: If you can write S as the span of a set of vectors, then you can simply apply Theorem 1.

(b) Let A be a m by n matrix. Let S denote the subset of R^m consisting of all the vectors $A\ x$, where x ranges over all the vectors in R^n. Prove that S is a subspace of R^m. See hint in (a).

```
> A := matrix([[0,-5,5],[-1,-5,4],[2,-3,5],[5,5,0]]);
>
```

$$A := \begin{bmatrix} 0 & -5 & 5 \\ -1 & -5 & 4 \\ 2 & -3 & 5 \\ 5 & 5 & 0 \end{bmatrix}$$

Student Workspace

```
>
```

Linear Algebra Modules Project
Chapter 5, Module 2

Basis and Dimension

Purpose of this module

The purpose of this module is to introduce the concepts of basis and dimension and to gain an understanding of them through both geometric and algebraic examples.

Prerequisites

Subspaces, linear independence, span, lines and planes, linear systems, matrix algebra.

Commands used in this module

```
> restart: with(linalg): with(lamp):
>
```

Tutorial

Section 1: Definition of Basis and Dimension

We all have an intuitive sense of the meaning of "dimension." We say that a line has dimension one (length), a plane has dimension two (length and width), and 3-space has dimension three (length, width, and height). For dimensions greater than 3, geometric intuition is less reliable and so our first task in this module will be to develop a general and unambiguous definition of dimension. Specifically, we will define the concept of the dimension of a subspace.

Consider these observations: A line through the origin is the span of a single vector, Span$\{u\}$, and a plane through the origin is the span of two vectors, Span$\{u, v\}$. Of course a plane can be the span of more than two vectors, but these would be linearly dependent vectors; we never need to use more than two vectors. Furthermore, R^3 is the span of three vectors, Span$\{u, v, w\}$. So it would seem that the dimension of a subspace is the fewest number of vectors needed to span that subspace and that these spanning vectors will always be linearly independent. This discussion leads us to make the following definitions:

```
>
```

Definition: Let S be any subspace of R^n. A set of vectors $\{v_1, v_2, ..., v_p\}$ in S is called a ***basis*** of S if S = Span$\{v_1, v_2, ..., v_p\}$ and $\{v_1, v_2, ..., v_p\}$ is linearly independent.

Definition: Let S be any subspace of R^n other than the zero subspace $\{0\}$. If S has a basis consisting of p vectors, then p is called the ***dimension*** of S. By convention, we say that the dimension of the zero subspace is 0.

Summary: A basis of a subspace S is a linearly independent set of vectors in S that spans S. The dimension of S is the number of vectors in a basis of S.

[>

Note: As we will see, a subspace can have many different bases, but all the bases for a given subspace have the same number of vectors. (See Theorem 4 in the Appendix to this module.) This assures us that the dimension of a subspace does not depend on which basis we choose for that subspace.

================

Example 1A: Let S = Span$\{u\}$, where $u = [1, 3, -2]$. So S is a line through the origin. Since $\{u\}$ is linearly independent and spans S, the dimension of S is 1 (the number of vectors in the basis $\{u\}$).

===============

===============

Example1B: Let S = Span$\{u, v\}$, where $u = [1, 3, -2]$ and $v = [2, -1, 0]$. So S is a plane through the origin. Since $\{ u, v \}$ is linearly independent and spans S, the dimension of S is 2 (the number of vectors in the basis $\{ u, v \}$).

===============

[>

Exercise 1.1: Let S = Span$\{ u, v, w \}$, where $u = [1, 3, -1, 3]$, $v = [2, -1, 0, -2]$ and $w = [4, 5, -2, 4]$.

(a) Check that $w = 2\,u + v$. Is the set of three vectors $\{ u, v, w \}$ a basis for S? Why or why not?

(b) Explain why Span$\{u,v,w\}$ = Span$\{u,v\}$.

(c) What is the dimension of S ?

Student Workspace

Answer 1.1

[>

===============

Example 1C: It should come as no surprise that the dimension of R^n is n. Here is the proof. The column vectors of the n by n identity matrix,

$$e_1 = \begin{bmatrix} 1 \\ 0 \\ . \\ . \\ . \\ 0 \end{bmatrix}, \quad e_2 = \begin{bmatrix} 0 \\ 1 \\ 0 \\ . \\ . \\ 0 \end{bmatrix}, \quad \ldots, \quad e_n = \begin{bmatrix} 0 \\ . \\ . \\ . \\ 0 \\ 1 \end{bmatrix},$$

[>

span R^n and are linearly independent. So $\{e_1, e_2, ..., e_n\}$ is a basis of R^n (called the "standard basis" of R^n). Since this set of vectors contains exactly n vectors, R^n has dimension n. For example, the standard basis of R^3 consists of the vectors

$$e_1 = \begin{bmatrix} 1 \\ 0 \\ 0 \end{bmatrix}, \quad e_2 = \begin{bmatrix} 0 \\ 1 \\ 0 \end{bmatrix}, \quad e_3 = \begin{bmatrix} 0 \\ 0 \\ 1 \end{bmatrix}$$

which are sometimes denoted i, j, k instead of e_1, e_2, e_3.

===============

[>

Section 2: How to Find a Basis

As we saw in Module 1, the two typical forms for a subspace are the solution set of a homogeneous matrix equation and the span of a set of vectors. We will see how to find a basis in each of these two circumstances.

===============

Example 2A: Solve the linear system $A\,x = 0$, where A is given below, and find a basis for the solution set. What is the dimension of the solution set?

```
[ > A := matrix([[-1,-1,-2,3,1],[-9,5,-4,-1,-5],[7,-5,2,3,5]]);
[ >
```

Solution: We use matsolve(..) to solve $A\,x = 0$ and then decompose the solution to find a basis:

```
[ > soln := matsolve(A,colvector([0,0,0]));
[ >
```

The decomposition of this solution is

$$x = \begin{bmatrix} -t_3 + t_2 \\ -t_3 + 2\,t_2 + t_1 \\ t_3 \\ t_2 \\ t_1 \end{bmatrix} = t_1 \begin{bmatrix} 0 \\ 1 \\ 0 \\ 0 \\ 1 \end{bmatrix} + t_2 \begin{bmatrix} 1 \\ 2 \\ 0 \\ 1 \\ 0 \end{bmatrix} + t_3 \begin{bmatrix} -1 \\ -1 \\ 1 \\ 0 \\ 0 \end{bmatrix} = t_1\,u + t_2\,v + t_3\,w$$

So the solution set is Span{u, v, w}. We must also check that { u, v, w } is a linearly independent set; that is, we must solve the equation $a\,u + b\,v + c\,w = 0$ and check that the only solution is when all the coefficients are zero. We rewrite the equation as $B\,x = 0$, where B is the matrix whose columns are the vectors u, v, w; then we solve $B\,x = 0$:

```
> u := colvector([0,1,0,0,1]):
  v := colvector([1,2,0,1,0]):
  w := colvector([-1,-1,1,0,0]):
> B := augment(u,v,w);
  zero := colvector(5,0);
> matsolve(B,zero);
>
```

Since the only solution to $B\,x = 0$ is the trivial solution $x = 0$, the set of vectors { u, v, w } is linearly independent. (In fact, we saw in Module 4 of Chapter 2 that whenever we decompose the solution set of a homogeneous linear system, the spanning vectors produced by the decomposition are always linearly independent. See Example 3A and Problem 9 of that module.) Therefore { u, v, w } is a basis for the solution set of $A\,x = 0$, and so the solution set has dimension 3 (the number of vectors in this basis).

==============

Next we consider a subspace that is given as a span of a set of vectors.

==============

Example 2B: Find a basis for S = Span{v_1, v_2, v_3, v_4} (see below). What is the dimension of this subspace?

```
> v1 := colvector([1,3,2,-5]):
  v2 := colvector([0,1,5,-3]):
  v3 := colvector([4,1,1,-1]):
  v4 := colvector([-2,5,3,-9]):
>
```

Solution:

(1) We begin by finding out whether the set {v_1, v_2, v_3, v_4} is linearly independent. If it is, then it is a basis for S and we're done. We construct a matrix A whose columns are the vectors v_1, v_2, v_3, v_4, and solve $A\,x = 0$. If {v_1, v_2, v_3, v_4} is linearly independent, this equation will have only the trivial solution.

```
> A := augment(v1,v2,v3,v4);
> soln := matsolve(A,colvector(4,0));
>
```

Since $A\,x = 0$ has nontrivial solutions, {v_1, v_2, v_3, v_4} is linearly dependent. Setting $t_1 = 1$ yields the solution [−2, 0, 1, 1] which in turn reveals the linear dependence relation $-2\,v_1 + v_3 + v_4 = 0$.

(2) We use the linear dependence relation to solve for v_3 in terms of v_1 and v_4: $v_3 = 2\,v_1 - v_4$. Since v_3 is just a linear combination of v_1 and v_4, we can eliminate it. In other words, Span{ v_1, v_2, v_4 } = Span { v_1, v_2, v_3, v_4 }.

(3) Is the subset consisting of only three vectors $\{v_1, v_2, v_4\}$ a basis for S? We once again check for linear independence.

```
> A := augment(v1,v2,v4):
  matsolve(A,colvector(4,0));
>
```

Since the set $\{ v_1, v_2, v_4 \}$ is linearly independent and spans S, it is a basis for S, which therefore has dimension 3.

==============

Exercise 2.1: Use the linear dependence relation that we found in Example 2B to find two other bases for the subspace S in Example 2B.

Student Workspace

Answer 2.1

```
>
```

Section 3: Coordinates Relative to a Basis

One way in which we will find a basis useful is that it gives us an alternative coordinate system. The usual perpendicular coordinate axes in the plane, for example, are related to the standard basis vectors $i = \begin{bmatrix} 1 \\ 0 \end{bmatrix}$ and $j = \begin{bmatrix} 0 \\ 1 \end{bmatrix}$, which are unit vectors that lie on the axes, point in the positive direction, and provide the unit of measurement. If we now have another basis $\{ u, v \}$ of R^2, we can think of this basis as determining an alternative coordinate system with axes that are not necessarily perpendicular and units of measurement on the two axes that may be different.

```
>
```

====================

Example 3A: The vector $w = \begin{bmatrix} 5 \\ 6 \end{bmatrix}$ can be written as a linear combination of the standard basis vectors i and j, $w = 5\,i + 6\,j$, where the coefficients are just the components of w. Let's consider another basis of R^2, $\{ u, v \}$, where $u = \begin{bmatrix} 3 \\ 1 \end{bmatrix}$ and $v = \begin{bmatrix} -1 \\ 4 \end{bmatrix}$. The vector w can also be written as a linear combination of u and v: $w = 2\,u + v$. The coefficients in this linear combination, 2 and 1, are no longer the components of w; however, we can think of them as "the coordinates of w relative to the basis $\{ u, v \}$."

Execute the basisgrid(..) command below to see an illustration of these coordinates. The picture shows the standard xy coordinate grid and the uv coordinate grid. The point [5, 6] with uv coordinates [2, 1] is shown as a small red circle.

```
> u := colvector([3,1]);
  v := colvector([-1,4]);
> basisgrid(u,v,[2,1]);
```

[>

=====================

Explore ~~ Enter other coordinates in the basisgrid(..) command and check the resulting picture. Can you see from the picture, for example, what the *xy* coordinates of the point with *uv* coordinates [−1, −2] are, and what the *uv* coordinates of the point with *xy* coordinates [−7, 2] are? Use basisgrid(..) to check your answers. Try similar explorations with the basis $\{\begin{bmatrix} -2 \\ -1 \end{bmatrix}, \begin{bmatrix} 1 \\ 3 \end{bmatrix}\}$.

[>

[>

Definition: Suppose $\boldsymbol{B} = \{ v_1, ..., v_p \}$ is a basis of a subspace S and w is a vector in S. Then the coefficients $c_1, ..., c_p$ in the linear combination $w = c_1\, v_1 + ... + c_p\, v_p$ are called the ***coordinates of w relative to the basis B***, or more succinctly, the ***B-coordinates of w***.

=====================

Example 3B: Find the coordinates of the vector w relative to the basis $\{ v_1, v_2, v_3 \}$ (see below).

```
> w := colvector([5,8,-10,2]);
> v1 := colvector([2,-3,0,4]):
  v2 := colvector([5,1,-2,0]):
  v3 := colvector([1,-1,2,-2]):
>
```

Solution: We must solve the equation $c_1\, v_1 + c_2\, v_2 + c_3\, v_3 = w$ for the coefficients c_1, c_2, c_3. This equation can be written $A\, x = w$, where A is the matrix whose columns are v_1, v_2, v_3, and x is the column vector of unknown coefficients. Thus:

```
> A := augment(v1,v2,v3);
> matsolve(A,w);
>
```

Therefore the coordinates of w relative to the given basis are [−1, 2, −3].

Question: How do we know in advance that the coordinates c_1, c_2, c_3 of w will be unique? Hint: Use the fact that $\{ v_1, v_2, v_3 \}$ is linearly independent.

=====================

Exercise 3.1: (a) Find the coordinates of the vector u below relative to the basis in Example 3B. (b) What are the coordinates of $3\,u$? What are the coordinates of $3\,u - 2\,w$? Try to answer these questions without a calculation.

```
> u := colvector([-10,-3,0,8]);
>
```

Student Workspace

Answer 3.1

[>

Appendix: Theorems on Basis and Dimension

The following theorems provide the fundamental properties of bases and dimension. Although we will not prove these theorems in any of the modules, your instructor may have you study the proofs in class or in a textbook. We will use these theorems freely.

Theorem 3: Suppose S = Span$\{v_1, v_2, ..., v_p\}$, and suppose $\{w_1, w_2, ..., w_q\}$ is another set of vectors in S. If $p < q$, then $\{w_1, w_2, ..., w_q\}$ is linearly dependent.

Theorem 4: If $\{v_1, v_2, ..., v_p\}$ and $\{w_1, w_2, ..., w_q\}$ are bases of a subspace S, then $p = q$.

Theorem 5: If S = Span$\{v_1, v_2, ..., v_p\}$ and S is not the zero subspace, then some subset of $\{v_1, v_2, ..., v_p\}$ is a basis of S.

Theorem 6: Suppose S is a subspace of dimension p. Then:
(a) Any set of more than p vectors in S is linearly dependent.
(b) Any set of fewer than p vectors in S will not span S.
(c) Any set of exactly p vectors in S is linearly independent if and only if it spans S.

Theorem 7: If S is a nonzero subspace of R^n, then S has a basis.

[>

Problems

Problem 1: Find bases by hand

Do all of this problem by hand.
(a) Find a basis for the solution set of $2x - y + 4z = 0$ (a plane). What is the dimension of the solution set?
(b) Find a basis for the solution set of the system of equations $\{2x - y + 4z = 0, x + 3z = 0\}$ (the line of intersection of two planes). What is the dimension of the solution set?

Student Workspace

Problem 2: Two different bases for one subspace

(a) Find a basis for the solution set of the homogeneous linear system below two different ways: Use matsolve(..) to solve $A\,x = 0$, where A is the coefficient matrix of the linear system, and use the solve(..) command. (Some of the commands you will need are supplied below.)
(b) Express each vector in the first basis as a linear combination of the vectors in the second basis, and express each vector in the second basis as a linear combination of the vectors in the first basis.
(c) In fact, whenever you have two different bases for one subspace, it is always possible to express

each vector in the first basis as a linear combination of the vectors in the second basis, and each vector in the second basis as a linear combination of the vectors in the first basis. Why?

```
> eqn1 := -2*x[1]+6*x[2]+4*x[3]-x[4]+x[5] = 0;
  eqn2 := 2*x[2]+2*x[3]-x[4]+x[5] = 0;
  eqn3 := 2*x[1]-x[2]-x[4]+x[5] = 0;
  eqn4 := x[1]+5*x[2]+7*x[3]-4*x[4]+4*x[5] = 0;
> A := genmatrix([eqn1,eqn2,eqn3,eqn4],
    [x[1],x[2],x[3],x[4],x[5]]);
> solve({eqn1,eqn2,eqn3,eqn4});
>
```

Student Workspace

Problem 3: Find a basis for the span of a set of vectors

(a) Find a basis for S = Span{ u, v, w, z }, where u, v, w, z are defined below. What is the dimension of this subspace S?

```
> u := colvector([-2, 3, -2, -1]);
  v := colvector([1, -1, 2, 2]);
  w := colvector([3, -3, -2, -4]);
  z := colvector([-3, 5, -2, 0]);
>
```

(b) Use the linear dependence relation you found in (a) to find all the subsets of { u, v, w, z } that are bases for S. For example, is { u, v, w, z } a basis of S? Is { u, v, w } a basis of S? { u, v, z }? { u, w, z }? { v, w, z }? { u, v }? (and so on). Explain.

Student Workspace

Problem 4: More than one linear dependence relation

(a) Find a basis for S = Span{v_1, v_2, v_3, v_4, v_5}, where v_1, v_2, v_3, v_4, v_5 are defined below. What is the dimension of S?

```
> v1 := colvector([2, 2, -5, 1, 2]):
  v2 := colvector([-5, 4, 2, -1, -7]):
  v3 := colvector([5, -1, -4, 2, 6]):
  v4 := colvector([-3, 0, 1, -2, -3]):
  v5 := colvector([0, 3, -2, 1, -1]):
>
```

(b) Use the linear dependence relations you found in (a) to find three subsets of {v_1, v_2, v_3, v_4, v_5} that are bases of S.

Student Workspace

Problem 5: Maximum dimension of a subspace

(a) Suppose S is the span of 4 vectors in R^5. What is the largest dimension that S could have? Explain.

(b) Suppose S is the span of 4 vectors in R^3. What is the largest dimension that S could have? Explain.

Student Workspace

Problem 6: Constructing new bases from old

(a) Let S = Span{ u, v, w }, where u, v, w are given below. Check that { u, v, w } is linearly independent (and hence is a basis of S).
(b) Is { $3u + 2v$, v, w } also a basis of S? For what values of the scalars a and b is { $a\,u + b\,v$, v, w } a basis of S? Explain geometrically.

```
> u := colvector([2, 4, 4, -3]);
  v := colvector([-4, 2, -4, 1]);
  w := colvector([4, 1, 1, -3]);
>
```

Student Workspace

Problem 7: Interchangeable bases?

Each of the sets of vectors { u_1, u_2 }, { v_1, v_2 }, { w_1, w_2 }, { z_1, z_2, z_3 } (see below) is a basis for a subspace of R^4. Two of these bases are bases for the same subspace of R^4. (a) Which two? (There's only one correct answer.) (b) Explain why your conclusion in (a) is correct by describing how the four subspaces are related to one another geometrically.

```
> u1 := colvector([2, 0, 4, -3]);
  u2 := colvector([-1, 2, -4, 1]);
> v1 := colvector([1,2,0,-2]);
  v2 := colvector([0,-4,4,1]);
> w1 := colvector([-1,2,-4,1]);
  w2 := colvector([-2,3,1,0]);
> z1 := colvector([1,2,0,-2]);
  z2 := colvector([3,2,4,-5]);
  z3 := colvector([-3,0,1,3]);
>
```

Student Workspace

Problem 8: Coordinates with respect to two bases

The set of vectors { u_1, u_2, u_3 } is a basis for a subspace S of R^4, and { v_1, v_2, v_3 } is a basis for the same subspace. A vector w in S has the coordinates [2, 4, −3] relative to the first basis. Find its coordinates relative to the second basis.

```
> u1 := colvector([1,-2,0,3]);
  u2 := colvector([0,-1,3,2]);
  u3 := colvector([2,-3,1,0]);
> v1 := colvector([1,-1,-3,1]);
  v2 := colvector([-1,1,1,1]);
  v3 := colvector([1,-2,2,1]);
```

[>

Student Workspace

Pencil and Paper Practice Questions

Note: Maple is not required for any of the problems in this section. Do all of these by hand.

1. Below are parametric equations for a line in R^3 through the origin. Find a basis for the line. What is the dimension of the line?

$$x = 2\,t, \qquad y = -5\,t, \qquad z = t$$

2. Find a basis for the solution set of the homogeneous linear equation below. (The coefficients of x_2 and x_3 are zero.) What is the dimension of the solution set?

$$x_1 - 2\,x_4 = 0$$

3. Find a basis for the solution set of the homogeneous linear system below. What is the dimension of the solution set?

$$x_1 - 2\,x_4 = 0$$
$$x_2 + x_3 = 0$$

4. If $\{u, v\}$ is a basis of a subspace S, is $\{2\,u, -3\,v\}$ a basis of S? Is $\{u + v, u - v\}$? Is $\{u + v, 2\,u + 2\,v\}$? Is $\{u, v, u + v\}$? Is $\{u + v\}$?

5. Can a vector in R^2 have the same coordinates relative to two different bases of R^2? If so, give an example; if not, explain why not.

Linear Algebra Modules Project
Chapter 5, Module 3

Subspaces Associated with a Matrix

Purpose of this module

The purpose of this module is to introduce the column space, row space, and null space of a matrix and to study the relationships among their dimensions. Also, the methods for finding bases that were introduced in Module 2, "Basis and Dimension," are explored in greater depth and additional methods are introduced.

Prerequisites

Basis and dimension of a subspace; matrix transformations; solution of linear systems.

```
[ >
```

Commands used in this module

```
[ > restart: with(linalg): with(lamp):
[ >
```

Tutorial

Section 1: Definitions and Introduction

Definition: Suppose A is an m by n matrix. Three important subspaces are associated with A:
The ***column space*** of A is the span of the column vectors of A, which is a subspace of R^m.
The ***row space*** of A is the span of the row vectors of A, which is a subspace of R^n.
The ***null space*** of A is the set of solutions of the homogeneous matrix equation $A\,x = 0$, which is a subspace of R^n.

```
[ >
```

================

Example 1A: Let A be the 2 by 3 matrix $\begin{bmatrix} 1 & 1 & 3 \\ 1 & 2 & 5 \end{bmatrix}$. Describe the three subspaces associated with this matrix, and find their dimensions.

The column space of A is the span of the column vectors $\begin{bmatrix} 1 \\ 1 \end{bmatrix}, \begin{bmatrix} 1 \\ 2 \end{bmatrix}, \begin{bmatrix} 3 \\ 5 \end{bmatrix}$. Note that these vectors are not a basis for the column space, since they are linearly dependent ($v_1 + 2\,v_2 = v_3$). However, any two of these three vectors can serve as a basis for the column space, which therefore has dimension 2. The column space is thus all of R^2.

The row space of A is the span of the row vectors [1, 1, 3] and [1, 2, 5]. Since these two vectors are linearly independent, they constitute a basis for the row space. Their span is therefore a subspace of dimension 2 (i.e., a plane through the origin) in R^3.

The null space of A is the solution set of the matrix equation $A\,x = 0$. We find the solution below.

```
[ > A := matrix([[1,1,3],[1,2,5]]);
[ > matsolve(A,colvector([0,0]));
[ >
```

So the null space consists of all multiples of the vector $\begin{bmatrix} -1 \\ -2 \\ 1 \end{bmatrix}$. This vector is a basis for the null space, which therefore has dimension 1. The null space is thus a line through the origin in R^3.

=================

Dimensions of the Column, Row, and Null Spaces

The dimensions of the column space, row space, and null space are closely related to one another. In this module, we will see why the relationships described in the next two theorems are true.

Theorem 8: The column space and row space of a matrix A have the same dimension. This dimension is equal to r, the number of pivots of A. (The dimension r is called the ***rank*** of A.)

Theorem 9: The dimension of the null space of a matrix A is $n - r$, where n is the number of columns of A and r is the rank of A.

```
[ >
```

For the 2 by 3 matrix A in Example 1A, we saw that both the column space and row space have dimension 2 (although one is a subspace of R^2 and the other is a subspace of R^3!). The null space, which is also a subspace of R^3, has dimension 1, and we observe that this dimension is equal to $n - r = 3 - 2$.

Exercise 1.1: Let $C = \begin{bmatrix} 1 & 2 & -1 \\ 2 & 4 & -2 \end{bmatrix}$. Find a basis for the column space of C, the row space of C, and the null space of C. Check that their dimensions agree with the theorems above. Describe each of the subspaces geometrically.

```
[ >
```

Student Workspace

Answer 1.1

Note: When referring to the rows of a matrix in Example 1A and Exercise 1.1, we naturally wrote the rows horizontally to match the way they appear in the matrix. However, we make no distinction between row vectors and column vectors as mathematical objects; for example, $[2, 3, -1]$ and $\begin{bmatrix} 2 \\ 3 \\ -1 \end{bmatrix}$ are considered to be the same vector in R^3. Normally we will write vectors vertically.

[>

Section 2: Geometric Examples

In this section, we study the column space and null space of matrix transformations. Recall that if A is a matrix, we define the corresponding matrix transformation T by $\mathrm{T}(x) = A\,x$. If A is a 2 by 2 matrix, then T is a function from R^2 to R^2. More generally, if A is an m by n matrix, then T is a function from R^n to R^m.

Some concepts associated with matrix transformations are analogous to concepts you have seen in your study of functions in precalculus and calculus. Consider, for example, $\mathrm{f}(x) = x^2 - 4$, which defines a function f from R to R. The set of "zeros" of such a function is the set of values of x such that $\mathrm{f}(x) = 0$; in this example, the set of zeros is $\{2, -2\}$. The "range" of such a function is the set of values of $\mathrm{f}(x)$, where x ranges over the domain R; in this example, the range is $\{y : -4 \le y\}$.

[>

We introduce now the analogous concepts for a matrix transformation T from R^n to R^m:

> ***Definition:*** The ***kernel*** of T is the set of vectors x in R^n such that $\mathrm{T}(x) = 0$; in other words, the kernel is the set of "zeros" of T.
>
> ***Definition:*** The ***range*** of T is the set of all values of $\mathrm{T}(x)$, where x ranges over all vectors in R^m; in other words, the range is the set of all output (or image) vectors of T in R^m.

Since $\mathrm{T}(x) = A\,x$, the kernel of the transformation T is equal to the null space of the matrix A. Also, the range of T is the set of all vectors $A\,x$, where x ranges over all the vectors in R^n. Now recall that $A\,x$ can be written as a linear combination of the columns of A where the weights are the components of x. Therefore the range of T is the set of all linear combinations of the columns of A, which is the column space of A. Thus:

> ***Theorem 10:*** Suppose A is an m by n matrix and T is the corresponding transformation from R^n to R^m defined by $\mathrm{T}(x) = A\,x$. Then (a) the kernel of T is the null space of A, and (b) the range of T is the column space of A.

[>

Note: In the examples below we will use the same letter for the matrix and for the corresponding matrix transformation, rather than use different letters such as A and T above.

================

Example 2A: Let P be the 2 by 2 projection matrix that projects vectors onto the x axis. Since P projects vectors that are on the y axis to 0, the kernel of the transformation P is the y axis and therefore the null space of the matrix P is also the y axis. The range of the transformation P is the x axis, and therefore the column space of the matrix P is also the x axis.

================

Note: We were able to solve Example 2A without referring to the numerical entries of the matrix P, and you will be asked to do the same in the exercises below. However, if you want to check your answers by looking at the numerical entries of the matrices, there are special commands for creating them, such as:

```
[ > projectmat(0);
[ > reflectmat(Pi/3);
[ > rotatemat(Pi/4);
[ >
```

Exercise 2.1: Let Q be the 2 by 2 projection matrix that projects vectors onto the line $y = x$. Reasoning geometrically as in Example 2A, find the null space and column space of Q. Again, you should be able to answer this question without referring to the numerical entries of the matrix Q.

Student Workspace

Answer 2.1

[>

Exercise 2.2:

(a) Let S be any 2 by 2 reflection matrix. Reasoning geometrically, find the null space and column space of S.

(b) Let T be any 2 by 2 rotation matrix. Reasoning geometrically, find the null space and column space of T.

(c) What are all the 2 by 2 matrices that have null space equal to $\{0\}$ and column space equal to R^2?

Student Workspace

Answer 2.2

[>

Exercise 2.3: Let P be the 3 by 3 projection matrix that projects vectors in R^3 onto the yz plane. So P applied to the vector $[x, y, z]$ results in the vector $[0, y, z]$. Reasoning geometrically, find the null space and column space of P.

Student Workspace

Answer 2.3

[>

Section 3: Method for Finding a Null Space Basis

In Example 2A of Module 2, "Basis and Dimension," you learned a method for finding a basis for the solution set of a homogeneous linear system $A\,x = 0$. Note that this solution set is also the null space of the matrix A. So let's review the method in Example 2A of Module 2:

================

Example 3A: Find a basis for the null space of the matrix A below:

```
[ > A := matrix([[-1,-1,-2,3,1],[-9,5,-4,-1,-5],[7,-5,2,3,5]]);
```

Solution: We use matsolve(..) to solve $A\,x = 0$ and then decompose the solution to find a basis:

```
[ > soln := matsolve(A,colvector([0,0,0]));
[ >
```

The decomposition of this solution is:

$$x = \begin{bmatrix} -t_3 + t_2 \\ -t_3 + 2\,t_2 + t_1 \\ t_3 \\ t_2 \\ t_1 \end{bmatrix} = t_1 \begin{bmatrix} 0 \\ 1 \\ 0 \\ 0 \\ 1 \end{bmatrix} + t_2 \begin{bmatrix} 1 \\ 2 \\ 0 \\ 1 \\ 0 \end{bmatrix} + t_3 \begin{bmatrix} -1 \\ -1 \\ 1 \\ 0 \\ 0 \end{bmatrix} = t_1\,u + t_2\,v + t_3\,w$$

So the null space of A is Span$\{u, v, w\}$, and $\{ u, v, w \}$ is a basis of the null space. Alternatively, we can use the nullbasis(..) command:

```
[ > nullbasis(A);
[ >
```

================

It is not entirely obvious that the spanning set in Example 3A, $\{ u, v, w \}$, is linearly independent. However, in Module 4 of Chapter 2, we saw that the set of vectors produced by the method of decomposition is always linearly independent. Here is the proof for the set of vectors $\{ u, v, w \}$ in Example 3A:

To show that $\{ u, v, w \}$ is linearly independent, we must show that the vector equation $t_1\,u + t_2\,v + t_3\,w = 0$ has only the trivial solution $t_1 = t_2 = t_3 = 0$. Let's rewrite the left side of this vector equation as a single vector:

$$t_1 \begin{bmatrix} 0 \\ 1 \\ 0 \\ 0 \\ 1 \end{bmatrix} + t_2 \begin{bmatrix} 1 \\ 2 \\ 0 \\ 1 \\ 0 \end{bmatrix} + t_3 \begin{bmatrix} -1 \\ -1 \\ 1 \\ 0 \\ 0 \end{bmatrix} = \begin{bmatrix} -t_3 + t_2 \\ -t_3 + 2\,t_2 + t_1 \\ t_3 \\ t_2 \\ t_1 \end{bmatrix} = \begin{bmatrix} 0 \\ 0 \\ 0 \\ 0 \\ 0 \end{bmatrix}$$

```
[ >
```

From the last three components of this equation, we see that $t_1 = t_2 = t_3 = 0$; therefore $\{ u, v, w \}$ is linearly independent. Note especially that the weights t_1, t_2, t_3 correspond precisely to the three free variables in the solution set of the linear system $A\,x = 0$. This explains why a similar proof works in the general case: as in this example, the weights in the linear combination correspond to the free variables in the solution set of $A\,x = 0$.

We can now see why the following theorem, which was first stated in Section 1, is true:

> ***Theorem 9:*** The dimension of the null space of a matrix A is $n - r$, where n is the number of columns of A and r is the rank of A.

```
>
```

Here's why. As we saw in the discussion above, the dimension of the null space of A equals the number of vectors produced by decomposing the solution set of $A\,x = 0$, and this number also equals the number of free variables in the solution set of $A\,x = 0$. Furthermore, in Theorem 3 of Chapter 1, we observed that the number of free variables in the solution set of $A\,x = 0$ is $n - r$, where n is the number of columns of A and r is the number of pivots of A. Therefore the dimension of the null space of A is $n - r$.

Exercise 3.1: (a) For the matrix B below, find rref(B) and use Theorem 9 to determine the dimension of the null space of B.
(b) Use matsolve(..) as in Example 3A to find a basis for the null space of B.

```
> B :=
  matrix([[2,2,-3,-8,5,-8],[-7,-7,-5,-3,-2,-3],[1,1,4,7,-3,7]]);
>
```

Student Workspace

Answer 3.1

Section 4: Method for Finding a Column Space Basis

In Module 2 you learned how to find a basis for a span of a set of vectors. Here we extend that method:

================

Example 4A: Find a basis for the column space of the matrix A in Example 3A:

```
> A := matrix([[-1,-1,-2,3,1],[-9,5,-4,-1,-5],[7,-5,2,3,5]]);
>
```

Solution: We first find a basis for the null space of A instead of the column space.

```
> nullbasis(A);
>
```

Each of these three null space vectors satisfies the matrix-vector equation $A\,x = 0$. Therefore, since the product $A\,x$ can be written as a linear combination of the columns of A, we get three equations of the form $x_1\,A_1 + ... + x_5\,A_5 = 0$. Recall that an equation of this form is called a "linear dependence relation." We therefore have found three linear dependence relations for the columns of A:

$$A_2 + A_5 = 0, \quad A_1 + 2\,A_2 + A_4 = 0, \quad -A_1 - A_2 + A_3 = 0$$

We solve for the rightmost vector in each of the above equations:

$$A_5 = -A_2, \quad A_4 = -A_1 - 2\,A_2, \quad A_3 = A_1 + A_2$$

[>

In each case, we expressed a column in terms of the columns A_1 and A_2. Therefore, $\{A_1, A_2\}$ spans the column space of A. Since $\{A_1, A_2\}$ is linearly independent, $\{A_1, A_2\}$ is a basis for the column space of A. So the basis we have found is:

$$\left\{ \begin{bmatrix} -1 \\ -9 \\ 7 \end{bmatrix}, \begin{bmatrix} -1 \\ 5 \\ -5 \end{bmatrix} \right\}$$

================

Convention: To apply the method of Example 4A systematically, we will always solve for the rightmost columns in each linear dependence relation.

Quick Method for Finding a Column Space Basis

For the matrix B below, we can find a column space basis very quickly.

```
[ > B := matrix([[1,0,0,-2],[0,1,0,1],[0,0,1,-1],[0,0,0,0]]);
[ >
```

Here's how: Let's denote the columns of B by B_1, B_2, B_3, B_4. Then $\{B_1, B_2, B_3\}$ (the set of pivot columns) is clearly a basis for the column space of B, and we can express the fourth column as a linear combination of the basis columns: $B_4 = -2\,B_1 + B_2 - B_3$.

This was quick because B is in reduced row echelon form, which suggests a quick method for finding a basis for the column space for any matrix A. Recall that every matrix A is row equivalent to a matrix R in reduced row echelon form, and that the equations $A\,x = 0$ and $R\,x = 0$ have the same solution set, since row operations do not change the solutions of a linear system. Therefore the columns of A and the columns of R satisfy the same linear dependence relations. Since the pivot columns of a matrix in reduced row echelon form are clearly a basis for the column space of that matrix, the corresponding columns of any row equivalent matrix are a basis for the column space of that matrix. We have thus shown:

[>

- The pivot columns of a matrix A constitute a basis for the column space of A. Therefore the dimension of the column space of A is the number of pivots of A.

Not only have we found a quick method for finding a column space basis, but we also proved part of Theorem 8! (See Section 1.)

====================

Example 4B: Use the quick method above to find a basis for the column space of the matrix A (repeated below) in Example 4A. Also, express each column of A as a linear combination of these basis columns.

```
> A := matrix([[-1,-1,-2,3,1],[-9,5,-4,-1,-5],[7,-5,2,3,5]]);
>
```

Solution:

```
> R := rref(A);
>
```

We see that the first and second columns of the reduced row echelon form R are the pivot columns of R; therefore the first and second columns of A constitute a basis for the column space of A. Furthermore, we see that the third column of R is the sum of the first two columns of R; therefore the same is true for the columns of A: $A_3 = A_1 + A_2$. Similarly, $A_4 = -A_1 - 2\,A_2$ and $A_5 = -A_2$.

Warning: Row operations change the column space of a matrix. For example, the matrices A and R above do not have the same column space, even though their columns satisfy the same linear dependence relations.

====================

Exercise 4.1: Use the quick method above to find a basis for the column space of the matrix N below. Also, express each column of N as a linear combination of these basis columns.

```
> N := matrix([[-4,-2,-3,3,-3],[-4,6,1,-1,1],
     [3,1,2,-2,3],[1,-1,0,0,1]]);
>
```

Student Workspace

Answer 4.1

Section 5: Method for Finding a Row Space Basis

Since the row space of A is also the column space of A^T, we could use the method of Section 4 to find a row space basis. However, rref(A) gives us a row space basis immediately:

================

Example 5A: Find a basis for the row space of the matrix B below:

```
> B := matrix([[-1,-2,-3,2],[4,-5,-1,5],[-2,2,0,-2]]);
> R := rref(B);
>
```

The nonzero rows of R form a basis for the row space of B:

$$\{[1, 0, 1, 0], [0, 1, 1, -1]\}$$

================

To see why the method of Example 5A works, we will explain two facts:

- Row operations do not change the row space of a matrix;
- The nonzero rows of a reduced row echelon form matrix are linearly independent.

Row operations merely replace rows of a matrix by linear combinations of the original rows. For example, let's denote the rows of the matrix B above by b_1, b_2, b_3. The first step in reducing matrix B to matrix R above might be to replace the second row of B by $4\ b_1 + b_2$. Thus, after one row operation, we have a new matrix C whose rows are b_1, c, b_3, where $c = 4\ b_1 + b_2$. These vectors are certainly in the row space of B. Also the rows of B are in the row space of C, since $b_2 = c - 4\ b_1$. A similar argument holds for any row operation, which shows that row operations do not change the row space of a matrix.

[>

One way to see that the nonzero rows of the reduced matrix R above are linearly independent is to look at just their entries in the pivot columns, which are columns 1 and 2. These are the standard basis vectors in R^2, [1, 0] and [0, 1], which we know are linearly independent. This argument will work for any reduced row echelon matrix: If you look at just the entries of the nonzero rows that are in the pivot columns, you will have the standard basis vectors, which we know are linearly independent.

We therefore conclude:

- The nonzero rows of the reduced row echelon form of a matrix A constitute a basis for the row space of A. Therefore the dimension of the row space of A is the number of nonzero rows of the reduced row echelon form of A.

[>

Since the number of nonzero rows of a row echelon form of A equals the number of pivots of A, we see from this conclusion and the similar conclusion in Section 4 that we have proved the other of our two main theorems of this module:

> ***Theorem 8:*** The column space and row space of a matrix A have the same dimension. This dimension is equal to r, the number of pivots of A.

Exercise 5.1: Find a row space basis for the matrix T below by the method of Example 5A.

```
> T := matrix([[-4,-4,3,1],[-2,6,1,-1],[-3,1,2,0],
  [3,-1,-2,0],[-3,1,3,1]]);
```

[>

⊞ Student Workspace

⊞ Answer 5.1

Problems

Problem 1 Find a null space basis

(a) Using the method of Example 3A, find a basis for the null space of the matrix A below. Check by using the nullbasis(..) command.
(b) What is the dimension of the null space of A? Use the definition of dimension to explain why your answer is correct.
(c) Compute rref(A) and use the result to confirm the dimension of the null space of A. Explain how it confirms the dimension. Hint: We can see the rank of the matrix from its echelon form.

```
> A := matrix([[-1,-2,-3,-2,-3],[-6,4,-2,-2,-5],
   [4,-4,0,2,3],[-1,-1,-2,3,0]]);
>
```

Student Workspace

Problem 2 Find a column space basis

(a) Using the method of Example 4A, find a basis for the column space of the matrix A in Problem 1 (repeated below).
(b) Write out the linear dependence relations that the null space vectors give you, and use them to write every column of A as a linear combination of the basis vectors in (a).
(c) Check your answer in (a) by computing rref(A). Explain what the reduced row echelon form tells you. Hint: Where are the pivots?

```
> A := matrix([[-1,-2,-3,-2,-3],[-6,4,-2,-2,-5],
   [4,-4,0,2,3],[-1,-1,-2,3,0]]);
>
```

Student Workspace

Problem 3: Projection example in R^3

Suppose P is the 3 by 3 matrix transformation that projects every vector in R^3 to the xy plane.
(a) (By hand) Reasoning geometrically, as in Section 2, find the null space and column space of P and their dimensions. Explain your reasoning.
(b) Find the matrix P. Hint: Use the fact that

$$P\begin{bmatrix} x \\ y \\ z \end{bmatrix} = \begin{bmatrix} x \\ y \\ 0 \end{bmatrix} \text{ for all points } \begin{bmatrix} x \\ y \\ z \end{bmatrix}$$

(c) Confirm your answers to (a) by applying appropriate Maple commands to the matrix P.

Student Workspace

Problem 4: Another projection example in R^3

Suppose Q is the 3 by 3 matrix that projects every vector in R^3 to the plane $x - 2y + 3z = 0$.
(a) (By hand) What is the null space of Q and what is the column space of Q? Find a basis for each of these two subspaces, and find their dimensions.
(b) Find the matrix Q as follows. Suppose the vectors you found in (a) are u, v, w. (You should have found three vectors, if you take the two bases together.) Then $Q\,u$, $Q\,v$, $Q\,w$ should be easy to figure out geometrically. Therefore you can write down the matrix product $Q\,[u, v, w]$ and then can solve an equation of the form $Q\,A = B$ to find Q.
(c) Confirm your answers to (a) by applying appropriate Maple commands to the matrix Q.

Student Workspace

Problem 5: Name that dimension (by hand)

Let A by an m by n matrix with r pivots. Express the following numbers in terms of m, n and r.
(a) The dimension of the column space of A.
(b) The dimension of the row space of A.
(c) The dimension of the null space of A.
(d) The dimension of the null space of A^T.
(e) The number of free variables in the solution set of $A\,x = 0$.
(f) The number of free variables in the solution set of $A^T\,x = 0$.

Student Workspace

Problem 6: Spaces associated with invertible matrices

If A is a n by n invertible matrix, what are the null space, column space, and row space of A? Explain.

Student Workspace

Problem 7: Two bases for a column space

(a) Find a basis for the column space of the matrix C below by the method of Section 4.
(b) Find a basis for the column space of C by the method of Example 5A applied to C^T.
(c) Just knowing that the bases you found in (a) and (b) are bases for the same subspace, how must the two sets of basis vectors be related to one another algebraically?

```
> C := matrix([[-2,0,2,1],[3,-1,-5,-3],[4,4,2,2],
     [4,-2,-6,-3],[2,0,-2,-1]]);
>
```

Student Workspace

Problem 8: True/False questions

For each of the following statements, say whether the statement is true or false, and justify your conclusion. If the statement is true, you might refer to a theorem that shows why it is true. If it is false, give a counter-example. If you are unsure whether a statement is true or false, you might use randmat(..) to construct some examples that you can test. For example, the command below constructs a random 5 by 3 matrix of rank 2.

```
[ > A := randmat(5,3,rank=2);
[ >
```

(a) If A is a 4 by 3 matrix, A must have a nonzero null space.

(b) If A is a 3 by 4 matrix, A must have a nonzero null space.

(c) If A is a 4 by 4 matrix whose null space is $\{0\}$, its column space must be all of R^4.

(d) If A is a 4 by 3 matrix whose null space is $\{0\}$, its column space must be all of R^4.

Student Workspace

Problem 9: Construct that matrix

For each of the following descriptions, either construct a matrix that satisfies the description or explain why no such matrix can exist.

(a) A is a 3 by 4 matrix of rank 4.

(b) A is a 3 by 4 matrix of rank 2.

(c) A is a 3 by 4 matrix of rank 3 and $A\begin{bmatrix}1\\2\\-1\\0\end{bmatrix}=0.$

(d) A is a 4 by 3 matrix of rank 3 and $A\begin{bmatrix}1\\2\\-1\end{bmatrix}=0..$

Student Workspace

Problem 10: Rank patterns for n by n matrices

For each of the following collections of n by n matrices, find the rank of all such matrices and justify your conclusion. For example, you might state a basis for the row space, column space, or null space of the matrices. You may use the special commands letterN(n), letterL(n), and bandmat(n) provided in the Student Workspace to construct and experiment with examples of these matrices. However, you may be able to see what the ranks must be without experimenting at all.

(a) The n by n matrices of 0's and 1's, where the 1's form the shape of the letter N.

(b) The n by n matrices of 0's and 1's, where the 1's form the shape of the letter L.

(c) The n by n matrices of 0's and 1's, where the 1's form two diagonal bands just above and below the main diagonal.

Student Workspace

Pencil and Paper Practice Questions

Note: Maple is not required for any of the problems in this section. Do all of these by hand.

1. Let P be the 2 by 2 projection matrix that projects onto the line $y = x$, and let R be the 2 by 2 rotation matrix that rotates vectors counterclockwise through the angle $\pi/2$. Answer the following questions by reasoning geometrically.
(a) What is the null space of the product matrix $R\,P$? of $P\,R$?
(b) What is the column space of the product matrix $R\,P$? of $P\,R$?

2. The solution set of the homogeneous matrix equation $A\,x = 0$ is given below:

$$\begin{bmatrix} \frac{31\,t_2}{4} + 6\,t_1 \\ \frac{1\,t_2}{4} \\ -\frac{5\,t_2}{2} - \frac{3\,t_1}{2} \\ t_2 \\ t_1 \end{bmatrix}$$

(a) Find a basis of the null space of A.
(b) From your answer to (a), determine the dimensions of the null space, column space, and row space of A.

3. Below is a matrix A and its reduced row echelon form R.

$$A = \begin{bmatrix} -1 & -2 & 5 & 4 \\ -8 & 3 & 2 & 0 \\ 2 & -3 & 4 & 5 \\ -6 & 3 & 0 & -3 \end{bmatrix} \qquad R = \begin{bmatrix} 1 & 0 & -1 & 0 \\ 0 & 1 & -2 & 0 \\ 0 & 0 & 0 & 1 \\ 0 & 0 & 0 & 0 \end{bmatrix}$$

(a) Find a basis for the column space of A, and express each column of A as a linear combination of these basis columns. (No calculations should be necessary.)
(b) From your answer to (a), determine the dimensions of the column space, row space, and null space of A.
(c) From your answer to (a), find two other bases of the column space of A that consist of columns of A.

Linear Algebra Modules Project
Chapter 5, Module 4

Loops and Spanning Trees -- An Application to Networks

Purpose of This Module

The purpose of this module is to provide a few examples of the ways in which bases arise and are used in practice.

Prerequisites

Null space and column space of a matrix; basis and dimension; solution of linear systems.

```
>
```

Commands used in this module

```
> restart; with(linalg): with(lamp):
>
```

Tutorial

Section 1: Traffic Flow Problem

The figure below shows traffic circulation and inflow/outflow from a two-block area with six intersections (called "nodes") and seven one-way streets (called "arcs") between them. The numbers and variables in the diagram represent the number of vehicles per hour that pass along each arc.

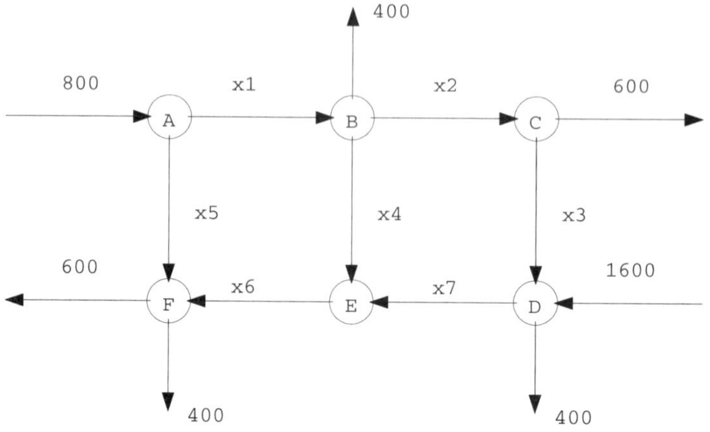


```
>
```

(The figure is taken from *Introduction to Linear Algebra, 3rd ed.*, by Johnson, Riess, and Arnold, Addison-Wesley, 1993, page 48.)

For example, every hour, 800 vehicles flow into Node A from outside the network (i.e., outside the two-block area) and must head toward Node B or Node F. The unknown traffic flow x_1 is the number of vehicles per hour that travel from Node A to Node B. (If x_1 is negative, the vehicles actually travel from B to A.)

We can find a system of linear equations satisfied by the seven unknown flows $x_1, ..., x_7$ by applying the "conservation of flow" principle:

- The number of vehicles that enter a node each hour must equal the number of vehicles that leave this node.

For example, the number of vehicles that enter Node A in an hour is 800, and the number of vehicles leaving A is $x_1 + x_5$; therefore:

```
> NodeA := x[1]+x[5]=800;
```

However, for reasons that will become clear later, we multiply the Node A equation by -1:

```
> NodeA := -x[1]-x[5]=-800;
>
```

The situation at Node B is more complicated: x_1 vehicles enter B, while 400, x_2, and x_4 vehicles exit B. We could therefore say that $x_1 = x_2 + x_4 + 400$; but, as we want all unknowns to be on the left side of the equation, we write:

```
> NodeB := x[1]-x[2]-x[4]=400;
```

Similarly, for the other four nodes we have:

```
> NodeC := x[2]-x[3]=600;
  NodeD := x[3]-x[7]=400-1600;
  NodeE := x[4]-x[6]+x[7]=0;
  NodeF := x[5]+x[6]=600+400;
>
```

Clearly, with six equations and seven unknown flows, there must be at least one free variable. So if our system of equations has any solution at all, it has infinitely many; that is, there are infinitely many flow patterns that accommodate the given inflows and outflows. We solve the system by writing it in the form $M\,x = b$, and discover that there are indeed solutions and that the system has two free variables:

```
> M := genmatrix([NodeA,NodeB,NodeC,NodeD,NodeE,NodeF],
   [x[1],x[2],x[3],x[4],x[5],x[6],x[7]]);
> b := colvector([-800,400,600,-1200,0,1000]);
> x := matsolve(M,b);
>
```

Since this solution set has two free variables, we can decompose the solution in the form

$$p + s\,u + t\,v$$

where p is a particular solution of the linear system, and $\{u, v\}$ is a basis for the solution set of the corresponding homogeneous system. Let's find a particular solution by setting both free variables to zero:

```
[ > p := subs({t[1]=0,t[2]=0},evalm(x));
[ >
```

Interpreting this particular solution is instructive: The values of t_2 (which is x_6), and t_1 (which is x_7) have been chosen to be zero, which forces x_4 to be zero; so no traffic passes through Node E in any direction. At Node D, on the other hand, there is a net inflow of $1600 - 400 = 1200$ vehicles, which flows backward ($x_3 = -1200$) to Node C. Then at Node C, 600 vehicles exit and the remaining 600 flow backward ($x_2 = -600$) to Node B . At Node B, 400 exit and the remaining 200 vehicles flow backward ($x_1 = -200$) to Node A. At Node A, 800 vehicles enter, joining the remaining 200, and the total flow forward ($x_5 = 1000$) to Node F. Finally, at Node F, the remaining $1000 = 600 + 400$ vehicles exit. So the flow is counterclockwise around the upper part of the loop DCBAF.

The basis vectors $\{u, v\}$ for the solution set of the corresponding homogeneous system $M\,x = 0$ are also interesting to interpret. Let's compute u and v:

```
[ > N := nullbasis(M);
[ > u := N[1];
[   v := N[2];
[ >
```

Keep in mind that these null space vectors represent flows where there is no traffic moving into or out of the network, since the right sides of the system of equations are all zero.

We now describe the traffic flow represented by the null space vector u. First note that, since $x_2 = x_3 = x_7 = 0$, no traffic flows through Nodes C or D. The first component, $x_1 = 1$, tells us that one vehicle goes from Node A to Node B; since $x_4 = 1$, this vehicle goes next to Node E; since $x_6 = 1$, this vehicle goes next to Node F; and since $x_5 = -1$, the vehicle returns to Node A. Summarizing, we can say that the vector u represents one vehicle travelling clockwise around the left-hand loop, ABEF.

```
[ >
```

Exercise 1.1: Describe similarly the traffic flow represented by the null space vector v.

Student Workspace

Answer 1.1

The remaining loop, ABCDEF, is discussed in Problem 1.

The two loop basis vectors u, v can give us useful insights about the solutions of our traffic flow example, since all solutions can be written in the form $p + s\,u + t\,v$, where p is the particular solution

```
[ > evalm(p);
[ >
```

Suppose, therefore, we decide to modify the flow represented by p as follows. Let's send another 1200 vehicles clockwise around the loop BCDE but no vehicles around the loop ABEF. That is, let's form the linear combination $p + 0\,u + 1200\,v$. Since p sends 1200 vehicles from Node D to Node C and 1200 v sends 1200 vehicles from Node C to Node D, the new value of x_3 will be zero. (So we can repair the potholes in that street.) Check this reasoning by executing the following command:

```
[ > evalm(p+1200*v);
[ >
```

Exercise 1.2: Find a solution that diverts all traffic from the arcs AB and BC, so that x_1 and x_2 are zero.

⊞ Student Workspace

⊞ Answer 1.2

Finally, let's examine the form of the coefficient matrix of our linear system:

```
[ > evalm(M);
[ >
```

Notice that each column consists of a single entry of -1 and a single entry of 1, with all remaining entries being 0. This is not accidental. The columns of M correspond to the unknowns x_1, ..., x_7, respectively, and the rows correspond to the nodes A, ..., F, respectively. So the entry -1 in the first column represents the fact that x_1 vehicles are leaving node A, and the entry 1 in the first column represents x_1 vehicles entering node B. We will see more matrices having this structure as we go on. (Aside: Earlier, when we multiplied the Node A equation by -1, we were helping to ensure this pattern of 1's and -1's. We can always obtain this pattern by consistently putting a minus sign before an unknown when it's an outflow and a plus sign before an unknown when it's an inflow.)

```
[ >
```

Section 2: An Oil Pipeline Construction Problem

The figure below shows a network consisting of an oil refinery at node 1 and oil fields at nodes 2 through 6. The arcs connecting them represent pipelines that could be constructed between them. Our goal in this problem is to find a smallest subset of arcs that would be sufficient to connect all the oil fields to the refinery.

The figure shows a flow of x_1 barrels of oil flowing from node 2 to node 1, x_2 barrels of oil flowing from node 4 to node 1, etc. In addition, imagine another arc coming out of node 1, which would represent the sum of the flows x_1, x_2, x_3 flowing into the refinery; this sum represents the total oil flow from all the oil fields. Also, at each of the nodes 2, 3, 4, 5, and 6, imagine oil flowing into that node from the corresponding oil field.

```
[ >
```

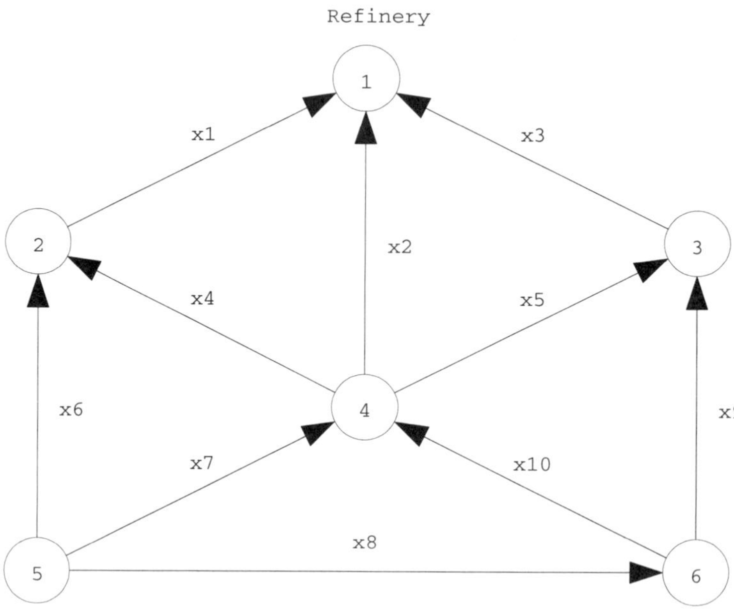

[>

If we had provided numerical values for the flows from each of the five oil fields, we could write, as we did in Section 1, six equations (one for each node) in the ten unknown flows, x_1, ..., x_{10}. However, we do have enough information to write down the coefficient matrix P of this system:

```
> P := matrix([[1,1,1,0,0,0,0,0,0,0],[-1,0,0,1,0,1,0,0,0,0],
  [0,0,-1,0,1,0,0,0,1,0],[0,-1,0,-1,-1,0,1,0,0,1],
  [0,0,0,0,0,-1,-1,-1,0,0],[0,0,0,0,0,0,0,1,-1,-1]]);
```

Note that the coefficient matrix P has the same structure as the coefficient matrix M in Section 1. For example, the first column of P has the entry -1 in the second row and the entry 1 in the first row; as in Section 1, this represents the flow x_1 going from node 2 to node 1.

[>

Exercise 2.1: Describe the meaning of column 4 of this matrix.

⊞ Student Workspace

⊞ Answer 2.1

[>

Now let's see how we can use the matrix P to find a smallest collection of arcs that will connect all the oil fields to the refinery. Notice that loops are undesirable, since they represent more than one route connecting two nodes. So we want to remove arcs that contribute to loops, but we do not want to remove arcs whose removal would disconnect the network. Removing an arc corresponds to removing a column of P. Therefore, since loops come from null space vectors (as we saw in Section 1), we want to remove columns of P until the null space of P becomes $\{0\}$. When the null space of P becomes $\{0\}$, the remaining columns will be linearly independent. However, if we remove a column of P that is not a linear combination of the remaining columns of P, removal of the corresponding arc would disconnect the network. So we want to remove only columns of P that are linear combinations of remaining columns of P. In other words, we want to remove columns of P until we are left with a basis of the column space of P!

```
> rref(P);
```

The pivots of P are in columns 1, 2, 3, 6, 8. So columns 1, 2, 3, 6, 8 of P form a basis for the column space of P. We construct the submatrix of P that has just these columns:

```
> Q := augment(column(P,1),column(P,2),column(P,3),
    column(P,6),column(P,8));
```

Since the columns of Q are linearly independent, the null space of Q is $\{0\}$. Let's check:

```
> matsolve(Q,colvector(6,0));
>
```

On your paper copy of the pipeline network, darken the five arcs corresponding to the five columns we chose above. Note the following facts about the smaller pipeline network you have drawn: Each node is connected to each of the other nodes, either directly by a single arc or by a sequence of adjoining arcs (so oil can flow from every oil field to the refinery); and no arc can be removed without cutting off at least one oil field from the refinery. Such a network is called a "spanning tree." So we have now designed a more efficient pipeline route. (Aside: A more realistic version of this problem would take into account the costs of constructing each of the pipelines, and we would seek a spanning tree that also minimizes the total construction costs.)

```
>
```

Problems

Problem 1: Another traffic loop

(a) Find a null space vector that represents the loop "one vehicle travelling counterclockwise around the two-block area FEDCBA."

(b) Express this vector as a linear combination of u and v, where $\{ u, v \}$ is the null space basis we found in Section 1.

Student Workspace

Problem 2: Another traffic diversion

Find a solution to the traffic flow example in Section 1 that diverts all traffic from the arcs AF and CD.

Student Workspace

Problem 3: Another spanning tree

By hand, draw a different spanning tree for our pipeline problem than the one we found in Section 2. Find the corresponding submatrix of P and check that its null space is $\{0\}$. Also check that its column space has the same dimension as the column space of P.

Student Workspace

Problem 4: Pipeline computation

(a) Solve the system of linear equations $P\,x = b$, where P is the pipeline matrix (repeated below) and b is the vector defined below. We can think of the first component of b as the total number of barrels of oil (in millions of barrels) that reach the refinery from all of the oil fields per year, and the remaining components as the number of barrels of oil that leave each of the oil fields.

```
> P := matrix([[1,1,1,0,0,0,0,0,0,0],[-1,0,0,1,0,1,0,0,0,0],
  [0,0,-1,0,1,0,0,0,1,0],[0,-1,0,-1,-1,0,1,0,0,1],
  [0,0,0,0,0,-1,-1,-1,0,0],[0,0,0,0,0,0,0,1,-1,-1]]);
> b := colvector([20,-2,-3,-1,-6,-8]);
>
```

(b) Write down the conservation of flow equation for node 2, and check that it corresponds to row 2 of the augmented matrix of the linear system $P\,x = b$.

(c) Decompose your answer in (a), as we did in our solution of the traffic flow problem in Section 1. On a drawing of the pipeline arcs, mark the loops corresponding to each of the null space basis vectors in your decomposition.

(d) Below is the matrix Q from Section 2 that corresponds to the spanning tree we found. Remove one of the columns of Q to form a new matrix M. By hand, draw the pipelines corresponding to the columns of M. Why is this not an acceptable solution to the pipeline problem? Also, solve $M\,x = b$ and explain why you get the result that you do.

```
> Q := augment(column(P,1),column(P,2),column(P,3),
   column(P,6),column(P,8));
>
```

Student Workspace

Problem 5: Dimensions, dimensions everywhere

Suppose E is an m by n matrix arising from a network of m nodes and n arcs in the same way as matrix M of Section 1 and matrix P of Section 2. So each column of E corresponds to one of the arcs and each row to one of the nodes; also, each column of E has one -1, one 1, and the remaining entries 0. (E is called the *incidence matrix* of the network.) Suppose also that E arises from a connected network. (A network is *connected* if each node can be joined to each of the other nodes by a sequence of adjoining arcs of the network.) It is not hard to see that any spanning tree for a connected network has exactly $m - 1$ arcs. (Two nodes are connected by one arc, three nodes by two arcs, etc.) After you read the following questions, you might want to perform some calculations on the additional examples below to see what's going on.

```
> E1 := matrix([[1,1,0],[-1,0,-1],[0,-1,1]]);
> E2 := matrix([[1,1,1,0,0,0],[-1,0,0,1,1,0],
   [0,-1,0,-1,0,-1],[0,0,-1,0,-1,1]]);
>
```

(a) What is the rank of the E? Explain.

(b) What is the dimension of the null space of E? Explain.

(c) What is the dimension of the null space of E^T? Find a basis for the null space of E^T.

Student Workspace

Pencil and Paper Practice Questions

Note: Maple is not required for any of the problems in this section. Do all of these by hand.

1. Draw a network corresponding to the incidence matrix below:

$$\begin{bmatrix} -1 & 0 & 0 & 1 \\ 1 & -1 & 0 & 0 \\ 0 & 1 & -1 & 0 \\ 0 & 0 & 1 & -1 \end{bmatrix}$$

2. Write down the 6 by 6 incidence matrix E for the network below. Then answer the following (preferably without performing any computations):
(a) What is the rank of E?
(b) Find a basis for the null space of E.
(c) Find a basis for the column space of E.

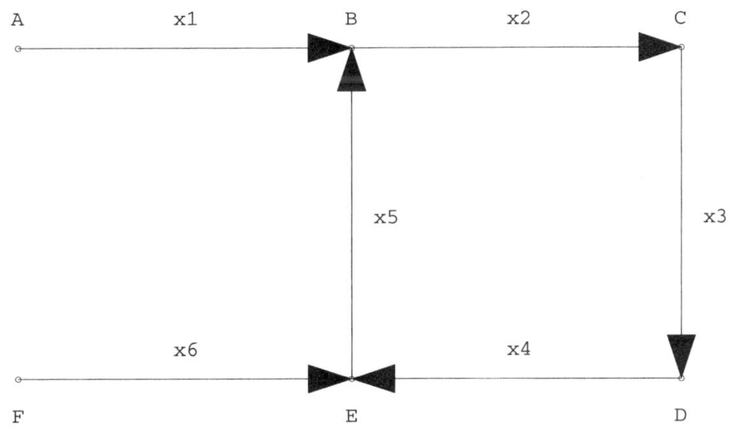

Linear Algebra Modules Project

Determinants

Purpose of this module

The purpose of this module is to discuss briefly the most important algebraic properties of determinants. Geometric properties of determinants are also illustrated.

Prerequisites

Familiarity with the Laplace cofactor expansion for calculating determinants. You should be able to use this expansion to calculate determinants of 2 by 2 and 3 by 3 matrices by hand.

Commands used in this module

```
> restart: with(plots): with(linalg): with(lamp):
>
```

Tutorial

Section 1: Row Operations and Determinants

In this section we outline important connections that exists between elementary row operations and determinants. We begin with a simple but extremely useful fact:

> ***Theorem 1:*** The determinant of an upper triangular matrix is the product of its diagonal entries.

```
> T := matrix([[4,5,6],[0,-2,7],[0,0,3]]);
> det(T);
>
```

In particular det(I) = 1, since the identity matrix I is diagonal and its diagonal entries are all 1's.

Exercise 1.1: (By hand) (a) Find the determinant of T by using Laplace cofactor expansions along the first column. (b) How would you use this technique to prove the above theorem?

Student Workspace

Answer 1.1

The next theorem describes the effect that row operations have on determinants:

> ***Theorem 2:*** If a matrix B is obtained from a matrix A by a row operation of the form:
> (a) $a\,R_i + R_j$ -> R_j (i.e., a multiple of one row is added to another), then $\det(B) = \det(A)$.
> (b) R_i <-> R_j (i.e., two rows are interchanged), then $\det(B) = (-1)\det(A)$.
> (c) $c\,R_i$ -> R_i (i.e., a row is multiplied by a scalar c), then $\det(B) = c\det(A)$.

This leads to a general technique for computing determinants without using Laplace cofactor expansions: Use row operations to reduce the matrix to upper triangular form, and then apply Theorem 1. Since we can reduce a matrix to upper triangular form using just the first two types of row operations, we will be changing the determinant by a sign, at most.

==================

Example 1A: Find the determinant of the following matrix by using row operations:

```
[ > A := matrix([[0,1,2],[1,-1,3],[2,0,1]]);
[ >
```

Solution: Our goal is to reduce A to echelon form and then apply Theorem 1 to find the determinant. As we go along, we will keep track of the changes in the determinant.
We begin by swapping the first and second rows.

```
[ > A1 := swaprow(A,1,2);
```

By Theorem 2(b) the determinant will change sign since we swapped rows; so $\det(A1) = -\det(A)$.
Next we add -2 times row 1 to row 3:

```
[ > A2 := addrow(A1,1,3,-2);
```

By Theorem 2(a), adding a multiple of one row to another doesn't change the determinant; so $\det(A2) = \det(A1)$. Adding -2 times row 2 to row 3 gets us to echelon form:

```
[ > A3 := addrow(A2,2,3,-2);
```

Again by Theorem 2(a), the determinant remains the same. Therefore

$$\det(A3) = \det(A2) = \det(A1) = -\det(A)$$

Now comes the easy part: By Theorem 1, we have $\det(A3)$ = (1) (1) (-9) = -9 . We therefore conclude that $\det(A) = 9$. Let's check our conclusion:

```
[ > det(A);
[ >
```

==================

Exercise 1.2: (By hand) Two matrices M and N are shown below.

$$M = \begin{bmatrix} 12 & 24 & 38 \\ -15 & -26 & -52 \\ 3 & 6 & 9 \end{bmatrix} \qquad N = \begin{bmatrix} 1 & 2 & 3 \\ 0 & 4 & -7 \\ 0 & 0 & 2 \end{bmatrix}$$

N is an echelon form for the matrix M. In fact, you can get from M to N by applying the following sequence of four row operations:

- 1. swap rows 1 and 3
- 2. mulitply row 1 by 1/3
- 3. add 15 times row 1 to row 2
- 4. add -12 times row 1 to row 3

Use this information and det(*N*) to determine det(*M*). You should be able to do this in your head.

Student Workspace

Answer 1.2

Determinants of Elementary Matrices

Recall that, for each of the three basic row operations, we can define a corresponding *elementary matrix E* that carries out the row operation by matrix multiplication. That is, if we multiply a matrix *A* on the left by *E*, then the product *E A* is the same as the row operation applied to *A*. Here is an example to refresh your memory:

===================

Example 1B: Suppose we want to swap the first two rows of the matrix *A* below. We can accomplish this by multiplying *A* on the left by the elementary matrix *E*, where *E* is obtained by applying the same row operation to the identity matrix.

```
> A := matrix([[0,1,2],[1,-1,3],[2,0,1]]);
> E := matrix([[0, 1, 0], [1, 0, 0], [0, 0, 1]]);
> EA := evalm(E&*A);
>
```

===================

Calculating the determinant of *E* is simple. Since the matrix *E* is itself the result of a swap row operation applied to the matrix *I*, we have by Theorem 2(b) that $\det(E) = (-1)\det(I)$ = -1. Similar conclusions can be reached for elementary matrices of the other two types. In summary:

- If *E* is an elementary matrix that adds a multiple of one row to another, then $\det(E) = \det(I)$, and therefore $\det(E) = 1$.
- If *E* is an elementary matrix that swaps two rows, then $\det(E) = -\det(I)$, and therefore $\det(E) = -1$.
- If *E* is an elementary matrix that multiplies a row by a scalar *c*, then $\det(E) = c\det(I)$, and therefore $\det(E) = c$.

Exercise 1.3: (By hand) Use Theorem 2 to show that if *A* is an *n* by *n* matrix and *E* is any *n* by *n* elementary matrix, then

- $$\det(E\,A) = \det(E)\,\det(A)$$

Student Workspace

Answer 1.3

[>

Section 2: Algebraic Properties of Determinants

These are the most important algebraic properties of determinants:

Theorem 3: If A and B are n by n matrices, then $\det(A\,B) = \det(A)\,\det(B)$.

Theorem 4: If A is an n by n matrix, then A is invertible if and only if $\det(A) \neq 0$.

Theorem 5: If A is an n by n matrix, then $\det(A^T) = \det(A)$.

Exercise 2.1: (By hand) Suppose that A and B are 4 by 4 matrices with $\det(A) = 5$ and $\det(B) = -4$. Calculate the following:
(a) $\det(A\,A^T)$
(b) $\det(A\,B^2\,A)$

Student Workspace

Answer 2.1

[>

Exercise 2.2: (By hand) Use Theorem 3 to prove the following (which is part of Theorem 4):

$$\text{If } A \text{ is invertible, then } \det(A) \neq 0 \text{ and } \det(A^{(-1)}) = \frac{1}{\det(A)}$$

Hint: Consider $\det(A^{(-1)}\,A)$.

Student Workspace

Answer 2.2

[>

Section 3: Geometric Properties of Determinants

The determinant of a matrix can be given a geometric interpretation.

==================

Example 3A: If A is a 2 by 2 matrix with row vectors u and v, then $|\det(A)|$ (the absolute value of the determinant of A) is equal to the area of the unit span of u and v (i.e., the parallelogram with u and v as two of its sides). Here is an example:

```
> u := [5,1];
> v := [1,4];
```

```
> A := matrix([u,v]);
> det(A);
>
```

So the area of the parallelogram below is 19.

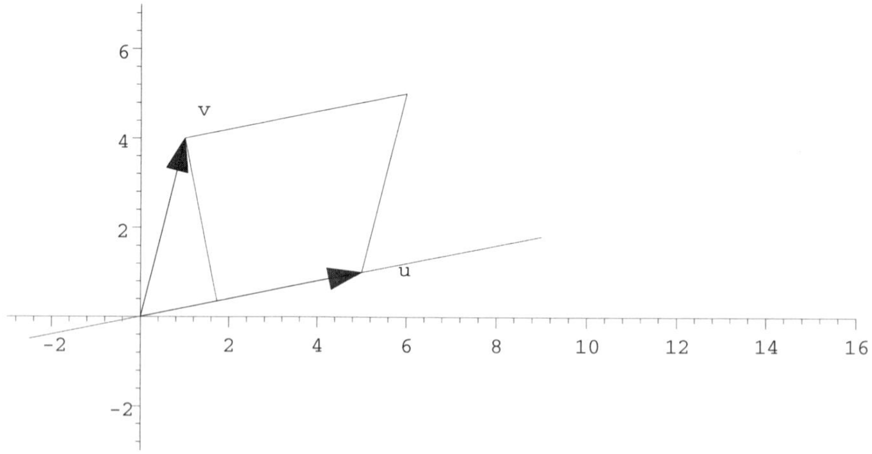


```
>
```

In the figure we have included the perpendicular line from the head of *v* onto the line containing *u*. The area of the parallelogram can be calculated from the formula $area = b\,h$, where *b* (base) is the length of the vector *u* and *h* (height) is the perpendicular distance between the base and the head of the vector *v*.

==================

Example 3B: In Section 1 of this module we observed that if we apply a row operation to a square matrix, we can predict exactly how its determinant will change. For example, by Theorem 2(a) we know if we start with a matrix *A* and add a multiple of one of its rows to another row, producing a second matrix *B*, then $\det(B) = \det(A)$. Our objective now is to justify this fact geometrically using what we know about determinants and area.

We begin with the matrix *A* and a second matrix *B* that is the result of adding 2 times the first row to the second row:

```
> A := matrix([u,v]);
> B := addrow(A,1,2,2);
>
```

The figure below shows the unit spans for both *A* (in white) and *B* (in yellow). The yellow parallelogram is the unit span for the vectors u and $w = 2\,u + v = [\,11, 6]$, where w is the second row of B. (Note: the vector w is shown with a blue head.)

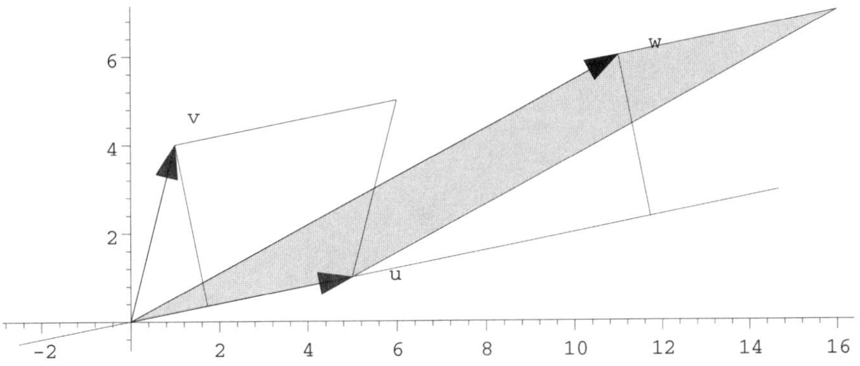

[>

Are the areas of the two parallelograms equal? Let's take the vector u as the base of the yellow parallelogram as well. Then notice that the perpendicular distance between the base and the head of w is the same as the distance from the base to the head of v, since the distance between two parallel lines is constant. So the product $b\,h$ remains the same for the second parallelogram. The areas are equal and hence the corresponding determinants are equal! So we now we see why $\det(B) = \det(A)$ when B is obtained from A by a row operation of the first type.

[>

===================

=====================

Example 3B: If A is a 3 by 3 matrix with row vectors u, v, and w, then $|\det(A)|$ (the absolute value of the determinant of A) is equal to the volume of the unit span of u, v, and w (i.e., the parallelepiped with u, v, and w as three of its sides. Here is an example:

```
> u := [4,0,3];
  v := [-1,5,1];
  w := [-2,3,7];
  A := matrix([u,v,w]);
> det(A);
```

So the volume of the unit span below is 149:

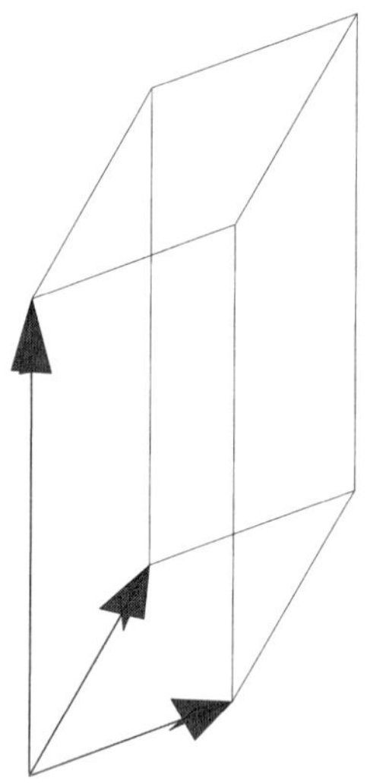

[>

===================

Note also that a row operation of the the first type changes only one face of the parallelogram and, by the geometric argument above, the area of this face is unchanged. So the volume is also unchanged.

Section 4: Appendix: Laplace Cofactor Expansion (a brief review)

The most important prerequisite for this module is familiarity with the Laplace cofactor expansion for calculating determinants. You should be able to use this expansion to calculate determinants of 2 by 2 and 3 by 3 matrices by hand.

Here is a brief review of the steps required in carrying out this expansion. Although we will use Maple's minor( ) command, that is merely to facilitate the discussion; you should be able to do this all by hand without the aid of Maple. In practice, calculation of determinants in Maple is accomplished with the det(..) command.

====================

Example 4A: Calculate the determinant of A using Laplace cofactor expansion across the first row.

```
[ > A := matrix([[1,3,2],[4,2,3],[1,3,5]]);
[ >
```

Solution: First define the "minors" for the first row. The minor for the entry in row i, column j is the 2 by 2 matrix that would remain if we crossed out the ith row and jth column.

```
[ > m11 := minor(A,1,1);
```

```
[ > m12 := minor(A,1,2);
[ > m13 := minor(A,1,3);
[ >
```

Now define the "cofactors" for the first row. The cofactor for the entry in row i, column j is $(-1)^{(i+j)}$ times the determinant of the minor for that entry.

```
[ > c11 := det(m11);
[ > c12 := (-1)*det(m12);
[ > c13 := det(m13);
[ >
```

Now compute the "cofactor expansion along the first row," which is the sum of the products of the entries of the first row and their cofactors:

```
[ > cofactor_expansion := (1)*c11+(3)*c12+(2)*c13;
[ >
```

We verify this result using Maple's det(..) command.

```
[ > det(A);
[ >
```

Exercise 4.1: (By hand) Use a cofactor expansion to find determinant of the matrix C.

$$C = \begin{bmatrix} 2 & 0 & 5 & 0 \\ 1 & -3 & 2 & 0 \\ 0 & 0 & 7 & 1 \\ 1 & 1 & 1 & 1 \end{bmatrix}$$

Student Workspace

Answer 4.1

```
[ >
```

Problems

Problem 1 : Laplace cofactor expansion

(By hand) Evaluate the determinants of the following two matrices by Laplace cofactor expansions.

```
[ > A := matrix([[-4,-4,3],[-4,-1,4],[1,-3,-1]]);
[ > B :=
    matrix([[0,-2,-2,2],[0,1,-1,-3],[-1,-1,1,2],[-1,-2,-3,-1]]);
[ >
```

Student Workspace

Problem 2 : Evaluating determinants by using row operations

Evaluate the determinants of the two matrices in Problem 1 (repeated below) by using row operations, as in Example 1A. You may use the swaprow(..) and addrow(..) commands.

```
[ > A := matrix([[-4,-4,3],[-4,-1,4],[1,-3,-1]]);
[ > B :=
```

```
matrix([[0,-2,-2,2],[0,1,-1,-3],[-1,-1,1,2],[-1,-2,-3,-1]]);
>
```

Student Workspace

Problem 3 : Algebra and determinants

If A and B are n by n matrices and c is a scalar, use the algebraic properties of determinants in Section 2 to prove each of the following identities:

(a) $\det(A\,B) = \det(B\,A)$.

(b) $\det(A^{(-1)}\,B\,A) = \det(B)$, if A is invertible.

(c) $\det(A^{(-1)}\,B\,A - c\,I) = \det(B - c\,I)$ if A is invertible.

(d) $\det(B^T - c\,I) = \det(B - c\,I)$.

Student Workspace

Problem 4 : Proof of det(AB)=det(A)det(B)

In Section 1, we saw that $\det(E\,B) = \det(E)\,\det(B)$ whenever E is an elementary matrix. Use this fact below.

(a) If $E1$ and $E2$ are both elementary matrices and A is their product, prove that $\det(A\,B) = \det(A)\,\det(B)$.

(b) In Module 6 of Chapter 3, we saw that every invertible matrix is the product of finitely many elementary matrices. Use this fact and the technique you used in part (a) of this problem to prove that $\det(A\,B) = \det(A)\,\det(B)$ whenever A is invertible.

Student Workspace

```
>
```

Linear Algebra Modules Project
Chapter 6, Module 1

Introduction to Eigenvalues and Eigenvectors

Purpose of this module

The purpose of this module is to introduce the concepts of eigenvector and eigenvalue from both algebraic and geometric points of view.

Prerequisites

Linear systems; null space of a matrix; basis and dimension; geometric transformations; determinants.

Commands used in this module

```
> restart; with(linalg): with(lamp):
>
```

Tutorial

Section 1: Definition and Examples of Eigenvectors and Eigenvalues

Definition: If A is a square matrix, we say that a nonzero vector x is an ***eigenvector*** of A and λ is the associated ***eigenvalue*** if the following equation holds:

$$A\,x = \lambda\,x$$

In other words, when x is multiplied by A, the result is just a scalar multiple of x.

==================

Example 1A: The vector $x = \begin{bmatrix} 3 \\ 2 \end{bmatrix}$ is an eigenvector of the matrix $A = \begin{bmatrix} 5 & -3 \\ -4 & 9 \end{bmatrix}$, since multiplying x by A results in a scalar multiple of x:

$$\begin{bmatrix} 5 & -3 \\ -4 & 9 \end{bmatrix}\begin{bmatrix} 3 \\ 2 \end{bmatrix} = \begin{bmatrix} 9 \\ 6 \end{bmatrix} = 3\begin{bmatrix} 3 \\ 2 \end{bmatrix}$$

So the eigenvalue associated with x is 3. Here is the same computation in Maple:

```
> A := matrix([[5,-3],[-4,9]]);
> x := colvector([3,2]);
> evalm(A&*x);
```

```
[ >
```

====================

Exercise 1.1: For the matrix *B* below, which of the vectors *u*, *v*, *w* are eigenvectors of *B*? For each one that is an eigenvector, what is its associated eigenvalue?

```
> B := matrix([[0,-1,1],[-2,-1,-1],[-2,1,-3]]);
> u := colvector([1, 2, 0]);
  v := colvector([-1, 1, 1]);
  w := colvector([1, 1, 1]);
[ >
```

Student Workspace

Answer 1.1

If x is an eigenvector of a matrix A with associated eigenvalue λ, then it turns out that every nonzero multiple of x will also be an eigenvector with the same eigenvalue. That is, if $A\,x = \lambda\,x$ then we also have $A\,w = \lambda\,w$ for every vector $w = c\,x$, where c is any nonzero scalar. Test this fact in the next example.

====================

Example 1B: In Example 1A, we saw that the vector $\begin{bmatrix} 3 \\ 2 \end{bmatrix}$ is an eigenvector of the matrix A with associated eigenvalue 3. Change the value of c (below) and observe that each nonzero multiple of x is also an eigenvector of A and that the associated eigenvalue is still 3.

```
> A := matrix([[5,-3],[-4,9]]);
> x := colvector([3,2]);
> c := 2;
  w := evalm(c*x);
  evalm(A&*w);
```

So $A\,w = 3\,w$.

```
[ >
```

====================

Exercise 1.2: (By hand) Use the rules of matrix algebra to prove the following statement:

- If A is a square matrix with eigenvector x and associated eigenvalue λ, then $A\,w = \lambda\,w$ for every vector $w = c\,x$, where c is any nonzero scalar.

Student Workspace

Answer 1.2

```
[ >
```

As we have seen, if we are given an eigenvector x of a matrix A, then finding its associated eigenvalue is simply a matter of comparing $A\,x$ with x. We now consider the reverse problem. If we are given a eigenvalue λ for a matrix A, how can we find a corresponding eigenvector x so that

$A\,x = \lambda\,x$? As the first step in solving the equation $A\,x = \lambda\,x$, we rewrite it in the form of a homogeneous matrix-vector equation:

$$A\,x = \lambda\,x$$
$$A\,x - \lambda\,I\,x = 0$$
$$(A - \lambda\,I)\,x = 0$$

Let's describe what we have just deduced:

> ***Theorem 1:*** A number λ is an eigenvalue of the square matrix A if and only if the equation $(A - \lambda\,I)\,x = 0$ has a nontrivial solution. The nontrivial solutions are the eigenvectors associated with λ.

====================

Example 1C: The matrix A defined below has an eigenvalue $\lambda = 3$. Use the above technique to find the associated eigenvectors.

```
[ > A := matrix([[1,2],[2,1]]);
[ >
```

Solution: We begin by calculating the matrix $A - 3\,I$. Note that this is simply the matrix A with 3 subtracted from each of its diagonal entries:

$$A - 3\,I = \begin{bmatrix} 1 & 2 \\ 2 & 1 \end{bmatrix} - \begin{bmatrix} 3 & 0 \\ 0 & 3 \end{bmatrix} = \begin{bmatrix} -2 & 2 \\ 2 & -2 \end{bmatrix}$$

We now solve the homogeneous linear system $(A - 3\,I)\,x = 0$, which we do using matsolve(..):

```
[ > M := evalm(A-3*identmat(2));
[ > soln := matsolve(M,colvector([0,0]));
[ >
```

Evidently, many eigenvectors are associated with this single eigenvalue, one for each nonzero value of t_1. Below we choose $t_1 = 1$ to find one such eigenvector and then check that it does work.

```
[ > x := subs(t[1]=1,evalm(soln));
[ > evalm(A&*x);
[ >
```

We can therefore describe the set of all eigenvectors associated with the eigenvalue 3 as the set of all nonzero multiples of the vector $\begin{bmatrix} 1 \\ 1 \end{bmatrix}$.

====================

Exercise 1.3: The matrix A in Example 1C has a second eigenvalue $\lambda = -1$. Find the set of eigenvectors associated with this eigenvalue. Also, show that $\lambda = 2$ is not an eigenvalue of A.

```
[ > A := matrix([[1,2],[2,1]]);
[ >
```

Student Workspace

Answer 1.3

Section 2: The Geometry of Eigenvectors

The definition of eigenvector has a simple geometric interpretation. If $A\,x = \lambda\,x$ then the vector $A\,x$ will point in the same or opposite direction as x (depending on whether $0 < \lambda$ or $\lambda < 0$), and the magnitude of the eigenvalue will correspond to the "stretch/shrink" factor. For example, recall that the vector $x = \begin{bmatrix} 3 \\ 2 \end{bmatrix}$ is an eigenvector of the matrix $A = \begin{bmatrix} 5 & -3 \\ -4 & 9 \end{bmatrix}$. Here is a picture of the vectors x and $A\,x$ drawn with red head and blue head respectively. Note that $A\,x$ is in the same direction as x but is 3 times as long.

```
> A := matrix([[5,-3],[-4,9]]):
  x := colvector([3,2]):
> drawvec2d(x,[A&*x,headcolor=blue],view=[-10..10,-10..10]);
> 
```

====================

Example 2A: Here is an example where we can find the eigenvalues and eigenvectors of a matrix simply by picturing how the matrix behaves as a geometric transformation. Let R be the matrix that reflects vectors across the line $y = x$:

$$R = \begin{bmatrix} 0 & 1 \\ 1 & 0 \end{bmatrix}$$

So R reflects vectors that lie along the line $y = -x$ to their negatives, and it keeps vectors that lie along the line $y = x$ fixed in place. Therefore every nonzero vector along the line $y = -x$ is an eigenvector with eigenvalue -1, and every nonzero vector along the line $y = x$ is an eigenvector with eigenvalue 1. Here is a picture of the eigenvector $u = \begin{bmatrix} -1 \\ 1 \end{bmatrix}$ (with a red head) and its image $R\,u = -1 \begin{bmatrix} -1 \\ 1 \end{bmatrix}$ (with a blue head):

```
> R := matrix([[0,1],[1,0]]):
> u := colvector([-1,1]):
  drawvec2d(u,[R&*u,headcolor=blue],view=[-2..2,-2..2]);
> 
```

====================

Even if we cannot figure out the eigenvalues of a matrix from our knowledge of its behavior as a geometric transformation, we will find pictures such as the one above quite helpful.

Let's consider the matrix A below as a matrix transformation. The vectors u and v below are eigenvectors of A with different eigenvalues. We will use drawvec2d(..) to draw the input vectors u and v with red heads and their corresponding output vectors with blue heads. See if you can guess the eigenvalue for each eigenvector just by looking at the picture.

```
> A := matrix([[1,1],[1/2,3/2]]);
```

```
> u := colvector([-2,1]);
> v := colvector([1,1]);
> drawvec2d(u,v,[A&*u,headcolor=blue],[A&*v,headcolor=blue]);
>
```

Sometimes we will find it easier to visualize eigenvectors by putting the tail of the output vector $A\,x$ at the head of the input vector x:

```
> drawvec2d([u,shaftcolor=red],[v,shaftcolor=red],
  [u,evalm(A&*u+u),shaftcolor=blue],
  [v,evalm(A&*v+v),shaftcolor=blue]);
>
```

Notice that the blue output vector in the second quadrant is half the length of the corresponding red input vector, and the other output vector is twice the length of its input. Hence the eigenvalues associated with u and v are 1/2 and 2, respectively.

In the next visualization, we show a large number of evenly-spaced input vectors with their corresponding output vectors. The input vectors are the red vectors of length 1 with tail at the origin, and each output vector has its tail at the head of the corresponding input vector and has a blue shaft. Can you spot the input vectors that are eigenvectors, and can you estimate their eigenvalues ? (Recall that if two vectors are parallel, they are multiples of one another.)

```
> eigpicture(A);
>
```

The next visualization is an animation that displays a rotating unit input vector (red) with its corresponding output vector (blue). Whenever the hands of this "clock" cross, we have an eigenvector with a positive eigenvalue. When they point in opposite directions, we have an eigenvector with a negative eigenvalue.

```
> eigclock(A);
>
```

Both the eigpicture(..) and eigclock(..) commands accept an optional second argument that allows you to specify the number of input vectors. For example, try eigpicture(A,24) .

In the next two exercises, you will use eigpicture(..) or eigclock(..) to investigate the eigenvectors and eigenvalues of some of the geometric transformations you studied in Module 1 of Chapter 4, "Geometry of Matrix Transformations of the Plane." Use the commands rotatemat(θ), reflectmat(θ), and projectmat(θ) to construct these transformations as you need them.

Exercise 2.1: Using eigpicture(..) and/or eigclock(..), experiment with the reflection matrix below, and then answer the following questions. What are the eigenvectors and eigenvalues of this reflection matrix? What are the eigenvectors and eigenvalues of every reflection matrix? (Try reasoning geometrically from your knowledge of what a reflection matrix does to vectors.)

```
> R := reflectmat(Pi/8);
>
```

Student Workspace

Answer 2.1

Exercise 2.2: Using eigpicture(..) and/or eigclock(..), experiment with the projection matrix below, and then answer the following questions. What are the eigenvectors and eigenvalues of this projection matrix? What are the eigenvectors and eigenvalues of every projection matrix? (Try reasoning geometrically from your knowledge of what a projection matrix does to vectors.)

```
[ > P := projectmat(Pi/4);
[ >
```

Student Workspace

Answer 2.2

```
[ >
```

Exercise 2.3: Let P be the projection matrix that projects vectors in R^3 onto the xy plane. P has two eigenvalues. Find them and their corresponding eigenvectors by using a geometric approach.

Student Workspace

Answer 2.3

Note: Recall that, by definition, the zero vector is never an eigenvector. However, the zero scalar can be an eigenvalue, as we saw in Exercises 2.2 and 2.3.

```
[ >
```

Section 3: The Characteristic Polynomial

In this section, we derive an equation for the eigenvalues of a matrix. This equation is derived by linking three key facts:

- λ is an eigenvalue of a matrix A if and only if the equation $(A - \lambda I)\,x = 0$ has a nontrivial solution. (This is Theorem 1 in Section 1.)
- If M is a square matrix, then the equation $M\,x = 0$ has a nontrivial solution if and only if M is singular (i.e., noninvertible). (This is part of Theorem 6 in Chapter 3.)
- A square matrix M is singular if and only if $\det(M) = 0$. (The expression $\det(M)$ denotes the determinant of M.)

```
[ >
```

Putting these together, we conclude:

Theorem 2: λ is an eigenvalue of a square matrix A if and only if $\det(A - \lambda I) = 0$.

====================

Example 3A: Use Theorem 2 to find the eigenvalues of the matrix A below.

```
[ > A := matrix([[1,2],[2,1]]);
```

Solution: We compute the expression $\det(A - \lambda I)$, where λ is treated as an unknown:

```
[ > M := evalm(A-lambda*identmat(2));
[ > p := det(M);
```

Then we solve the quadratic polynomial equation det($A - \lambda I$) = 0 for λ:

```
[ > solve(p=0);
[ >
```

So the eigenvalues of A are -1 and 3.

====================

Definition: The polynomial p(λ) = det($A - \lambda I$) is called the ***characteristic polynomial*** for the matrix A. The equation p(λ) = 0 is called the ***characteristic equation*** of A. (Its roots are the eigenvalues of A.)

A theorem from elementary algebra says that if r is a root of a polynomial equation p(λ) = 0, then $\lambda - r$ is a factor of the polynomial p(λ). So if we factor the characteristic polynomial p(λ) in the form

$$p(\lambda) = (\lambda - r_1)(\lambda - r_2) \ldots (\lambda - r_n)$$

then $r_1, r_2, \ldots, r_n$ are the roots of p(λ) and hence are the eigenvalues of A.

Definition: The number of times that a factor $\lambda - r$ occurs in the characteristic polynomial is called the ***algebraic multiplicity*** of the eigenvalue r.

Let's factor the characteristic polynomial in Example 3A to find the algebraic multiplicities of the two eigenvalues:

```
[ > factor(p);
[ >
```

Since the factor $\lambda + 1$ occurs just once, -1 has algebraic multiplicity 1; similarly, 3 has algebraic multiplicity 1. We can confirm our above results by using Maple's command charpoly(..), which computes the characteristic polynomial of a matrix.

```
[ > charpoly(A,lambda);
[ >
```

Maple note: The command charpoly(A,lambda) actually computes det($\lambda I - A$) instead of det($A - \lambda I$); so if A has an odd number of rows and columns, the polynomial produced by charpoly(..) will have opposite sign from the polynomial we get using det($A - \lambda I$).

We can also confirm the results of Example 3A by using the command eigenvals(..), which computes all the eigenvalues at once. It also shows the algebraic multiplicity of the eigenvalues by repeating each eigenvalue the number of times it appears in the factorization of the characteristic polynomial:

```
[ > eigenvals(A);
```

Since each eigenvalue appears just once in this list, each has algebraic multiplicity 1.

```
[ >
```

Exercise 3.1: In answering the questions about the matrix A below, do not use charpoly(..) or eigenvals(..) except as a check on your answers.

```
> A := matrix([[4,10,-10],[-5,-11,5],[-5,-5,-1]]);
>
```

(a) Find the characteristic polynomial of A.

(b) Factor the characteristic polynomial, and use the factorization to determine the eigenvalues of A and their algebraic multiplicities.

Student Workspace

Answer 3.1

====================

Example 3B: Finding the eigenvalues of an upper triangular matrix is particularly easy, since the determinant of an upper triangular matrix is simply the product of its diagonal entries. Here, for example, is an upper-triangular matrix K and the matrix $K - \lambda I$:

$$K = \begin{bmatrix} 2 & -3 & 1 \\ 0 & -4 & 5 \\ 0 & 0 & 7 \end{bmatrix} \qquad K - \lambda I = \begin{bmatrix} 2-\lambda & -3 & 1 \\ 0 & -4-\lambda & 5 \\ 0 & 0 & 7-\lambda \end{bmatrix}$$

So $\det(K - \lambda I) = (2-\lambda)(-4-\lambda)(7-\lambda)$, which equals zero when λ is 2 or -4 or 7. So the eigenvalues of an upper-triangular matrix are its diagonal entries. We confirm:

```
> K := matrix([[2,-3,1],[0,-4,5],[0,0,7]]):
> eigenvals(K);
>
```

====================

Degree of the Characteristic Polynomial

As you may have observed by now, the characteristic polynomial of a 2 by 2 matrix has degree 2 and the characteristic polynomial of a 3 by 3 matrix has degree 3. More generally:

- The characteristic polynomial of an n by n matrix has degree n.

Since a polynomial of degree n has at most n roots, we conclude:

Theorem 3: An n by n matrix has at most n distinct eigenvalues.

Caution: The characteristic equation $p(\lambda) = 0$ is especially handy for finding eigenvalues of 2 by 2 matrices. But, since the degree of $p(\lambda)$ gets larger as the matrices get larger, the equation $p(\lambda) = 0$ may become difficult or even impossible to solve for larger matrices. Maple can sometimes help here by solving the equation for us. Furthermore, if we and Maple are unable to compute the eigenvalues exactly, Maple can approximate them numerically. However, as we will see in the next section, there are often better ways to find eigenvalues and eigenvectors.

```
>
```

Section 4: Eigenspaces and Geometric Multiplicity

In Section 1, we saw that to find the eigenvectors associated with a given eigenvalue λ of a matrix A, we needed to solve the equation $(A - \lambda I)\,x = 0$. The set of solutions of this equation is the null space of the matrix $A - \lambda I$, which is the subject of the next definition:

> ***Definition:*** If A is a square matrix with eigenvalue λ, then the null space of the matrix $A - \lambda I$ is called the ***eigenspace*** associated with the eigenvalue λ.

Note that the eigenspace associated with λ includes the zero vector, although the zero vector is not an eigenvector. All the other vectors in the eigenspace, however, are eigenvectors associated with λ.

```
[ >
```

> ***Definition:*** The dimension of the eigenspace associated with an eigenvalue λ is called the ***geometric multiplicity*** of the eigenvalue λ. (That is, the dimension of the null space of $A - \lambda I$ is the geometric multiplicity of λ.)

We will see that the algebraic and geometric multiplicity of an eigenvalue are often, but not always, the same.

====================

Example 4A: For the matrix A below, find a basis for the eigenspace associated with the eigenvalue $\lambda = -6$. What is the geometric multiplicity of this eigenvalue? What is its algebraic multiplicity?

```
[ > A := matrix([[4,10,-10],[-5,-11,5],[-5,-5,-1]]);
```

Solution: We solve the equation $A\,x = -6\,x$ by writing it in the form $(A + 6\,I)\,x = 0$:

```
[ > soln := matsolve(A+6*identmat(3),colvector([0,0,0]));
[ >
```

To find a basis of this eigenspace, we decompose the solution in the usual way:

$$\begin{bmatrix} -t_2 + t_1 \\ t_2 \\ t_1 \end{bmatrix} = t_1 \begin{bmatrix} 1 \\ 0 \\ 1 \end{bmatrix} + t_2 \begin{bmatrix} -1 \\ 1 \\ 0 \end{bmatrix}$$

Or we can use the nullbasis(..) command to find a basis for the null space in one step:

```
[ > nullbasis(A+6*identmat(3));
```

Since the eigenspace basis has two vectors, the geometric multiplicity of the eigenvalue -6 is 2. The characteristic polynomial below shows that -6 also has algebraic multiplicity 2.

```
[ > factor(charpoly(A,t));
[ >
```

====================

====================

Example 4B: Here's an example where we can figure out the eigenvalues and eigenspaces by purely geometric reasoning. Find the eigenvalues and corresponding eigenspaces for the 3 by 3 matrix P that projects vectors in R^3 onto the plane $2x - y + 3z = 0$.

Solution: If v is any vector in this plane, P projects v to itself: $Pv = v$. Thus, every nonzero vector in this plane is an eigenvector of P with eigenvalue 1. The geometric multiplicity of 1 is the dimension of the plane, which is 2.

The vectors orthogonal to the plane $2x - y + 3z = 0$, which are all the multiples of the normal vector $[2, -1, 3]$, are projected by P to the zero vector; thus, $Pv = 0\,v$ for all these vectors. So 0 is an eigenvalue of P, and the dimension of its associated eigenspace is 1.

====================

```
[ >
```

Finding Eigenvalues and Eigenspaces by Recognizing Patterns

Recall that λ is an eigenvalue of the matrix A if and only if the matrix $A - \lambda I$ is singular (Theorem 1). So, if we can discover by careful observation of patterns those values of λ which make $A - \lambda I$ singular, we will have found the eigenvalues of A. Note that each of the following conditions is equivalent to the condition that $A - \lambda I$ is singular (see Theorem 6 of Chapter 3 and Theorem 9 of Chapter 5):

- the columns of $A - \lambda I$ are linearly dependent.
- the rows of $A - \lambda I$ are linearly dependent.
- the rank of $A - \lambda I$ is less than n (where A is n by n).
- the null space of $A - \lambda I$ is nonzero.

These conditions will often help us discover quite quickly some or all of the eigenvalues of A.

====================

Example 4C: Consider the matrix A below and the matrix $A - \lambda I$:

```
> A := matrix([[4, 4, 0], [4, 4, 0], [1, 2, 1]]);
  evalm(A-lambda*identmat(3));
[ >
```

Observe that when $\lambda = 1$, the last column of $A - \lambda I$ is zero, and hence the columns of $A - 1\,I$ are linearly dependent. Therefore $\lambda = 1$ is an eigenvalue of A.

====================

Exercise 4.1: Find the remaining two eigenvalues of the matrix A above by observing those values of λ which make the columns or rows of A linearly dependent.

Student Workspace

Answer 4.1

[>

We can also find the geometric multiplicity of the eigenvalues of the matrix A above by a simple observation rather than by a calculation. This observation is based on Theorem 9 of Chapter 5, which says:

- The dimension of the null space of an n by n matrix of rank r is $n - r$.

For the eigenvalue $\lambda = 1$ of the matrix A above, observe that the first two columns of $A - 1\ I$ are linearly independent. Therefore the rank of $A - 1\ I$ is 2, and hence the dimension of the null space of $A - 1\ I$ is $3 - 2 = 1$. This shows that $\lambda = 1$ has geometric multiplicity 1. By a similar argument, the other two eigenvalues of A also have geometric multiplicity 1.

[>

More generally, since the geometric multiplicity of an eigenvalue λ of a matrix is the dimension of the null space of $A - \lambda\ I$, we have:

> ***Theorem 4:*** The geometric multiplicity of the eigenvalue λ of an n by n matrix A is $n - r$, where r is the rank of $A - \lambda\ I$.

====================

Example 4D: Use Theorem 4 to find the geometric multiplicity of the eigenvalue $\lambda = 2$ of the matrix A below.

```
> A := matrix([[2,0,0,1],[1,2,0,1],[1,0,2,1],[1,0,0,2]]);
  evalm(A-lambda*identmat(4));
```

[>

Solution: The matrix $A - 2\ I$ has two zero columns, and therefore is singular; hence 2 is an eigenvalue of A. Furthermore, since the first and fourth columns of $A - 2\ I$ are linearly independent, the rank of $A - 2\ I$ must be 2. Therefore the eigenvalue 2 has geometric multiplicity $4 - 2 = 2$. Calculation of a null space basis for $A - 2\ I$ confirms this conclusion:

```
> nullbasis(A-2*identmat(4));
```

[>

====================

Exercise 4.2: Find another eigenvalue of the matrix A above, and find its geometric multiplicity.

```
> A := matrix([[2,0,0,1],[1,2,0,1],[1,0,2,1],[1,0,0,2]]):
  evalm(A-lambda*identmat(4));
```

[>

Student Workspace

Answer 4.2

We state the next theorem without proof:

> ***Theorem 5:*** The geometric multiplicity of an eigenvalue is less than or equal to its algebraic multiplicity.

Since the algebraic multplicities of all the eigenvalues of an n by n matrix cannot add up to more than the degree n of the characteristic polynomial, we conclude from Theorem 5:

> ***Theorem 6:*** The sum of the geometric multiplicities of the eigenvalues of an n by n matrix is less than or equal to n.

Exercise 4.3: We have seen that the 4 by 4 matrix A of Example 4D (repeated below) has eigenvalue 2 with geometric multiplicity 2 and eigenvalue 1 with geometric multiplicity 1. What more can we infer about this matrix from Theorem 6?

```
> A := matrix([[2,0,0,1],[1,2,0,1],[1,0,2,1],[1,0,0,2]]);
>
```

Student Workspace

Answer 4.3

Diagonal Matrices

====================

Example 4E: The eigenvalues and eigenspaces of a diagonal matrix are especially easy to find. Let M be the diagonal matrix defined below. Observe what happens when we multiply each of the standard basis vectors of R^3 by M. Use the results to find the eigenvalues and eigenvectors for M.

```
> M := diag(7,4,-3);
> e1 := colvector([1,0,0]);
  evalm(M&*e1);
> e2 := colvector([0,1,0]);
  evalm(M&*e2);
> e3 := colvector([0,0,1]);
  evalm(M&*e3);
>
```

Therefore the standard basis vector e_1 is an eigenvector with eigenvalue 7, e_2 is an eigenvector with eigenvalue 4, and e_3 is an eigenvector with eigenvalue -3.

====================

Exercise 4.4: Guess the eigenvalues and the geometric multiplicities of the following 4 by 4 diagonal matrix K . Check your answer by having Maple find a basis for each of the associated eigenspaces.

```
> K := diag(3,5,-2,5);
>
```

Student Workspace

Answer 4.4

Exercise 4.5: Based on the examples so far, state a general rule for determining the eigenvalues of any diagonal matrix and their geometric multiplicity.

Student Workspace

Answer 4.5

```
>
```

Section 5: Eigenvects(..) Command

The powerful command eigenvects(A) gives you the eigenvalues of *A*, the algebraic multiplicity of each eigenvalue, and a basis for each eigenspace. However, some care is required to interpret the output correctly.

====================

Example 5A: Interpret the results of the command eigenvects(A) below.

```
> A := matrix([[3,0,0,1],[1,3,0,1],[1,0,3,1],[1,0,0,3]]);
> E := eigenvects(A);
>
```

Solution: The output of the eigenvects(..) command is a sequence of lists, where each list in the sequence has three components. The output will be easier to interpret if we look at each list in the sequence separately:

```
> E[1]; E[2]; E[3];
>
```

Each of the above lists provides three pieces of information about just one eigenvalue. The first piece of information is the eigenvalue itself; the second is its algebraic multiplicity; and the third is a basis of its eigenspace. For example, 4 is an eigenvalue of A with algebraic multiplicity 1, and { [1, 2, 2, 1] } is a basis of its eigenspace. Similarly, 3 is an eigenvalue with algebraic multiplicity 2, and { [0, 1, 0, 0], [0, 0, 1, 0] } is a basis of its eigenspace.

====================

Warning: If you apply the eigenvects(..) command repeatedly to the same matrix, Maple may produce output that changes from one application of eigenvects(..) to another. That is, it may produce the eigenvalues in a different order and may produce a different basis for some of the eigenspaces.

Exercise 5.1: What is the remaining eigenvalue of the matrix *A* in Example 5A, and what are its algebraic and geometric multiplicities? (Determine your answers from the output above.)

Student Workspace

Answer 5.1

```
>
```

Problems

Problem 1: Find eigenvectors from eigenvalues

(a) The matrix A below has the eigenvalue $\lambda = -4$. Find all the associated eigenvectors.
(b) The columns of A are linearly dependent. What eigenvalue must A therefore have? Find all the eigenvectors associated with this eigenvalue.

```
> A := matrix([[-3, -1, 2], [1, -5, 2], [2, -2, 0]]);
>
```

Student Workspace

Problem 2: Work backwards from eigenvalues and eigenvectors

(a) Find a 2 by 2 matrix A, that has eigenvalues 2 and 3 with associated eigenvectors [1,4] and [-2,3], respectively. (Hint: Write the equations that A must satisfy; then rewrite these equations in the form of a single matrix equation $A\,B = C$, where B and C are known 2 by 2 matrices.
(b) Now consider the general problem: Find a 2 by 2 matrix A with eigenvalues λ_1 and λ_2 and associated eigenvectors v_1 and v_2, respectively, where $\{ v_1, v_2 \}$ is linearly independent. Will this problem always have a solution? Explain.

Student Workspace

Problem 3: Stochastic matrices

The matrix M below is a Markov matrix we encountered in Module 4 of Chapter 3. Recall that the state vectors $x_k = M^k\, x_0$ converge to a limit vector x (called the steady-state vector) and that $M\,x = x$.

So the steady-state vector $x = \begin{bmatrix} .6 \\ .4 \end{bmatrix}$ is an eigenvector with associated eigenvalue 1:

```
> M := matrix([[.94,.09],[.06,.91]]);
> x := colvector([.6,.4]);
> evalm(M&*x);
```

(a) Use the eigenvals(..) command to compute the eigenvalues of the matrix M and the other stochastic matrices below.

```
> N := matrix([[.9,.03,.09],[.02,.91,.04],[.08,.06,.87]]);
> W := matrix([[1,.5,0,0],[0,0,.5,0],[0,.5,0,0],[0,0,.5,1]]);
> Z := matrix([[0,.5,0,0],[1,0,.5,0],[0,.5,0,1],[0,0,.5,0]]);
>
```

(b) What patterns do you see in the eigenvalues of these stochastic matrices, in addition to the fact that 1 is an eigenvalue?
(c) Prove that 1 is an eigenvalue of every stochastic matrix. Hint: What nonzero vector is in the null space of $M^T - I$?

Student Workspace

Problem 4: Rotation matrices in R^2

(a) Let M be the 2 by 2 matrix that rotates vectors through an angle θ in the counterclockwise direction. Use geometry to explain why most rotation matrices have no eigenvalues or eigenvectors. To stimulate your geometric insight, use eigclock(..) on one or two specific rotation matrices.
(b) A few rotation matrices, however, do have eigenvalues and eigenvectors. Which ones are these? What are the eigenvalues and their corresponding eigenvectors?
(c) Apply the eigenvals(..) command to the general rotation matrix

$$M = \text{rotatemat}(\theta) = \begin{bmatrix} \cos(\theta) & -\sin(\theta) \\ \sin(\theta) & \cos(\theta) \end{bmatrix}$$

Notice that there are real eigenvalues for certain values of θ only. What are these values of θ and what eigenvalues do they produce? Compare this answer to your answer in (b).

Student Workspace

Problem 5: Shear matrices

Let S be the horizontal shear matrix:

$$S = \text{xshearmat(k)} = \begin{bmatrix} 1 & k \\ 0 & 1 \end{bmatrix}$$

(a) (By hand) Find the eigenvalues of S from the characteristic polynomial of S. Also find the algebraic and geometric multiplicities of each eigenvalue.
(b) Give a geometric interpretation of the eigenvalues of S and their associated eigenspaces.

Student Workspace

Problem 6: Matrix transformations of R^3

By geometric reasoning alone, determine the eigenvalues and eigenvectors of the 3 by 3 matrix transformations below. You may perform calculations on the sample matrices in the Student Workspace to stimulate your intuition or confirm your answers.

(a) Let R be the 3 by 3 matrix that reflects vectors in R^3 across a plane through the origin, $a\,x + b\,y + c\,z = 0$. What are the eigenvalues of R and their associated eigenspaces?

(b) Let M be the 3 by 3 matrix that rotates vectors in R^3 about a line through the origin, Span$\{ v \}$, through the angle θ (where θ is not a multiple of π). This matrix has one real eigenvalue. What is this eigenvalue and its associated eigenspace?

(c) Let P be the 3 by 3 matrix that projects vectors in R^3onto a line through the origin, Span$\{v\}$. This matrix has two eigenvalues. What are the eigenvalues of P and their associated eigenspaces?

Student Workspace

Problem 7: Eigenvalues of related matrices

Once we find the eigenvectors and eigenvalues of a matrix A, we can deduce the eigenvectors and eigenvalues of a great many other matrices that are derived from A. For example, as you will see, the matrices A^2, A^3, etc. have the same eigenvectors as A.

(a) What is the relation between the eigenvalues of A^n (n any positive integer) and those of A? (You might calculate the eigenvalues and eigenvectors of the matrices below and some of their powers to help you spot the pattern.)

(b) What is the relation between the eigenvalues and eigenvectors of $A + k\,I$ (k any real scalar) and those of A? (Again, some calculations may help you spot the pattern.)

(c) (By hand) Prove your statement in (a) for A^2 and your statement in (b) for $A + 2\,I$. Hint: Since A and A^2 have the same eigenvectors, see what happens when you multiply an eigenvector of A by A^2:

$$A^2\,x = A\,A\,x = ?$$

```
> A := matrix([[1,2],[2,1]]);
> B := matrix([[1,1],[1/2,3/2]]);
> C := projectmat(Pi/4);
>
```

Student Workspace

Problem 8: Nilpotent matrices

A matrix A is said to be *nilpotent* if $A^n = \mathrm{O}$ for some positive integer n. Check that the matrices A and B below are nilpotent.

(a) For each matrix, find the eigenvalues. (You are not required to find their algebraic or geometric multiplicities.)

(b) What are the eigenvalues for every nilpotent matrix? Explain why your answer is correct. Hint: Problem 6 is helpful.

```
> A := matrix([[1,-1,2],[3,-3,6],[1,-1,2]]);
> B := matrix([[1,0,0,-1],[1,1,0,-4],[0,1,1,-6],[0,0,1,-3]]);
>
```

Student Workspace

Problem 9: Practice using the characteristic polynomial

For the matrices A and B below, do all of the following (without using charpoly(..) or eigenvals(..) or eigenvects(..), except as a check):

(a) Find the characteristic polynomial and factor it.

(b) Find the eigenvalues and their algebraic multiplicities.

(c) Find a basis for each eigenspace and the geometric multiplicity of each eigenvalue.

```
> A := matrix([[4,-1,1],[-1,4,-1],[1,-1,4]]);
> B := matrix([[-1,-1,2,-3],[4,3,0,6],[0,0,5,9],[0,0,-1,-1]]);
>
```

Student Workspace

Problem 10: Matrices of ones

(a) Use the commands below to construct the n by n matrix J of all 1's. What are the eigenvalues of J and their geometric multiplicities? (Suggestions: Examine the matrix $J - \lambda\,I$ to look for patterns that suggest values of λ for which $J - \lambda I$ is singular, and keep in mind that the eigenvalues can depend on n. If that doesn't help enough, use the eigenvals(..) and nullbasis(..) commands or the eigenvects(..) command.)

(b) For every n by n matrix of all 1's, explain why one of the eigenvalues must always have an eigenspace of dimension $n - 1$. (Hint: What can you say about the rank of the matrix $A - \lambda I$ for this λ?)

(c) (By hand) What are the eigenvalues and their geometric multiplicities for the n by n matrix that has all 1's, except for -2's down the main diagonal? (Hint: This matrix is closely related to the matrix of all 1's.)

```
> n := 4:   # change the value of n to experiment.
  J := matrix(n,n,1);
  evalm(J-lambda*identmat(n));
> 
```

Student Workspace

Problem 11: Letter N matrices

A "letter N matrix" is a square matrix of 0's and 1's, where the 1's form the shape of the letter N. Use the commands below to construct and study letter N matrices of various sizes.

(a) What are the eigenvalues and their geometric multiplicities for the n by n letter N matrix? (See the suggestions for Problem 10(a).)

(b) For every n by n letter N matrix, explain why one of the eigenvalues must always have an eigenspace of dimension $n - 2$.

(c) State a basis for the eigenspace associated with each of the eigenvalues other than the eigenvalue in part (b).

```
> n := 4:   # change the value of n to experiment.
> N := letterN(n);
  evalm(N-lambda*identmat(n));
> 
```

Student Workspace

Problem 12: Letter L matrices

A "letter L matrix" is a square matrix of 0's and 1's, where the 1's form the shape of the letter L. Use the commands below to construct and study letter L matrices of various sizes.

(a) What are the eigenvalues and their geometric multiplicities for the n by n letter L matrix? (See the suggestions for Problem 10(a).)

(b) For every n by n letter L matrix, explain why the eigenvalues have the geometric multiplicities claimed in (a).

```
> n := 4;   # change the value of n to experiment.
> L := letterL(n);
  evalm(L-lambda*identmat(n));
> 
```

Student Workspace

Pencil and Paper Practice Questions

Note: Maple is not required for any of these problems. Do all of these by hand.

1. Consider the matrix

$$A = \begin{bmatrix} 8 & 6 & 2 \\ -10 & -8 & -2 \\ 5 & 3 & -1 \end{bmatrix}$$

(a) Find the eigenvalue associated with the eigenvector $x = \begin{bmatrix} 2 \\ -2 \\ 1 \end{bmatrix}$.

(b) Find a basis for the eigenspace associated with the eigenvalue -2. What is the geometric multiplicity of -2?

2. Find the characteristic polynomial for the reflection matrix $\begin{bmatrix} \cos(2\,\theta) & \sin(2\,\theta) \\ \sin(2\,\theta) & -\cos(2\,\theta) \end{bmatrix}$ and solve the characteristic equation to find the eigenvalues. Are these the eigenvalues predicted by geometric reasoning?

3. Let R be the 3 by 3 matrix transformation that reflects every vector across the plane $2\,x - y + 3\,z = 0$. What are the eigenvalues of R and their associated eigenspaces? (Do not find R; just reason geometrically.)

Linear Algebra Modules Project
Chapter 6, Module 2

Eigenvector Analysis of Discrete Dynamical Systems

Purpose of this module

The purpose of this module is to apply the concepts of eigenvectors and eigenvalues to the analysis of sequences x_k defined by equations of the form $x_{k+1} = A\,x_k$.

Prerequisites

Definition and basic properties of eigenvalues and eigenvectors.

Commands used in this module

```
> restart; with(linalg): with(plots): with(lamp):
>
```

Tutorial

Section 1: Eigenvector Bases for R^n

If an n by n matrix A has n linearly independent eigenvectors, we can use these vectors to form an *eigenvector basis* for R^n. For example, the 2 by 2 matrix A below has eigenvalues 36 and 18 with linearly independent eigenvectors $u = \begin{bmatrix} 4 \\ 3 \end{bmatrix}$ and $v = \begin{bmatrix} -2 \\ 3 \end{bmatrix}$ respectively. Therefore the set $\boldsymbol{B} = \{\,u, v\,\}$ is a basis for R^2 that consists entirely of eigenvectors; hence, it is an eigenvector basis.

```
> A := matrix([[30,8],[9,24]]);
> u := colvector([4,3]);
  v := colvector([-2,3]);
> evalm(A&*u);
  evalm(A&*v);
>
```

====================

Example 1A: Express the vector $y = \begin{bmatrix} 6 \\ 9 \end{bmatrix}$ as a linear combination of the eigenvectors u and v.

Solution: We are looking for a_1 and a_2 such that $a_1\,u + a_2\,v = y$. This can be rewritten as the matrix-vector equation $P\,a = y$, where $P = [\,u\ v\,]$ and $a = \begin{bmatrix} a_1 \\ a_2 \end{bmatrix}$. Since P is invertible (why?), we can solve for a: $a = P^{(-1)} y$. We calculate a_1 and a_2 below.

```
> y := colvector([6,9]);
  P := augment(u,v);
> a := evalm(P^(-1)&*y);
```

Therefore $a_1 = 2$ and $a_2 = 1$. Let's check that these values of a_1 and a_2 do yield the given vector y:

```
> a1 := 2;
  a2 := 1;
> evalm(a1*u+a2*v);
>
```

We say that the point $y = [6, 9]$ has *coordinates* $[2, 1]$ *relative to the basis* ***B***. So y has standard coordinates $[6, 9]$ and ***B****-coordinates* $[2, 1]$. (See discussion of coordinates relative to a basis in Section 3 of Module 2 of Chapter 5.)

====================

Execute the next input line to see a picture that shows both coordinate systems, the standard *xy* grid and the *uv* coordinate grid. Notice that the point with ***B***-coordinates $[2, 1]$ (shown with a red circle) is the point $y = [6, 9]$.

```
> basisgrid(u,v,[2,1]);
>
```

Exercise 1.1: (a) From the above picture, estimate the ***B***-coordinates of the point $y = [5, 6]$. Test your answer by changing the ***B***-coordinates in the basisgrid(..) command above.

(b) Calculate the ***B***-coordinates exactly by having Maple find $P^{(-1)} y$ for $y = \begin{bmatrix} 5 \\ 6 \end{bmatrix}$.

Student Workspace

Answer 1.1

Summary: If y is a point in standard coordinates and x is that same point in ***B***-coordinates, we can switch between the two by the conversion formulas

$$x = P^{(-1)} y \qquad \text{and} \qquad y = P\,x$$

where P is the matrix whose columns are the vectors in the basis ***B***. Note that while we have been using this approach in the context of an eigenvector basis, it applies more generally to any situation where we have a basis for R^n other than the standard basis. See Problem 1 for an example.

```
>
```

Section 2: Migration Model -- Lampland Revisited

We begin this section by revisiting the urban versus rural population problem we discussed in the Markov Chain module. Recall that we started off with the following assumptions about Lampland:

- (1) Each year 6% of the urban population moves to rural areas and the remaining 94% stay in urban areas.

- (2) Each year 9% of the rural population moves to urban areas and the remaining 91% stay in rural areas.
- (3) At the beginning of our study, the Lampland population is 45% urban and 55% rural.

We then set up a 2 by 2 migration matrix M and a sequence of state vectors, $x_0, x_1, x_2, \ldots$, that were related by the recursion equation

$$x_{k+1} = M\,x_k$$

where:

```
> x0 := colvector([.45,.55]);
  M := matrix([[.94,.09],[.06,.91]]);
>
```

We also found that the sequence of states could be calculated more easily by using powers of the matrix M. Specifically, we found that $x_k = M^k\,x_0$ for every value of k.

Eigenvector decomposition

Maple note: In the computations below, we will display only four significant digits in the answers, since the remaining digits in this particular example are all zeros. The command Digits:=n sets the number of digits to be displayed at n; the default value of Digits is 10.

We now investigate a different method for computing the iterates x_k, one based on the eigenvalues and eigenvectors of M. First we find the eigenvalues and eigenvectors of M:

```
> Digits := 4;
> eigenvals(M);
  Id := identmat(2):
> nullbasis(M-Id);
> nullbasis(M-.85*Id);
>
```

So $v_1 = \begin{bmatrix} 1.5 \\ 1 \end{bmatrix}$ is an eigenvector of M with eigenvalue $\lambda_1 = 1$, and $v_2 = \begin{bmatrix} -1 \\ 1 \end{bmatrix}$ is an eigenvector with eigenvalue $\lambda_2 = .85$. Since v_1 and v_2 are linearly independent, $\{v_1, v_2\}$ is an eigenvector basis of R^2. So we can express the initial state x_0 as a linear combination of v_1 and v_2:

$$x_0 = a_1\,v_1 + a_2\,v_2$$

From this we can find x_k:

$$x_k = M^k\,x_0 = M^k\,(a_1\,v_1 + a_2\,v_2) = a_1\,M^k\,v_1 + a_2\,M^k\,v_2$$

Now recall the fact that if v is an eigenvector of a matrix A with eigenvalue λ so that $A\,v = \lambda\,v$, then

$A^k v = \lambda^k v$ for every positive integer k. (Problem 7 in Module 1). Therefore

$$x_k = a_1 \lambda_1^k v_1 + a_2 \lambda_2^k v_2$$

[>

With little effort, the above derivation can be modified to prove the following eigenvector decomposition theorem for n by n matrices:

***Theorem 7*:** Let A be an n by n matrix with n linearly independent eigenvectors $v_1, ..., v_n$ and associated eigenvalues $\lambda_1, ..., \lambda_n$. Then, for any vector x_0 in R^n, we have the eigenvector decomposition

$$A^k x_0 = a_1 \lambda_1^k v_1 + ... + a_n \lambda_n^k v_n$$

where the scalars $a_1, ..., a_n$ are determined by the equation $x_0 = a_1 v_1 + ... + a_n v_n$.

[>

====================

Example 2A: Use the above eigenvector decomposition to find the steady-state vector for the Lampland urban/rural population problem and to plot the urban and rural populations as functions of the time parameter k.

Solution: First we list again the eigenvalues and associated eigenvectors of the migration matrix M:

```
> v1 := colvector([1.5,1]);
  v2 := colvector([-1,1]);
> t1 := 1;
  t2 := .85;
```

Next we find the weights a_1, a_2 in the equation $a_1 v_1 + a_2 v_2 = x_0$ by solving $P a = x_0$ for a, where $P = [v_1, v_2]$and $a = \begin{bmatrix} a_1 \\ a_2 \end{bmatrix}$:

```
> evalm(x0);
  P := augment(v1,v2);
  a := evalm(P^(-1)&*x0);
```

[>

Therefore $a_1 = .4$ and $a_2 = .15$, and so the eigenvector decomposition formula in Theorem 7 is:

$$x_k = .4\ 1^k \begin{bmatrix} 1.5 \\ 1 \end{bmatrix} + .15\ .85^k \begin{bmatrix} -1 \\ 1 \end{bmatrix}$$

Notice that as k becomes large, the factor $.85^k$ approaches 0, and thus the iterates x_k get closer and closer to the steady-state vector $.4\begin{bmatrix}1.5\\1\end{bmatrix}=\begin{bmatrix}.6\\.4\end{bmatrix}$. This method for computing the iterates x_k provides insights that our earlier methods did not; it shows clearly how each eigenvector and eigenvalue affects the long-term behavior of x_k. The decomposition also provides us with an explicit formula for each component of the vector x_k:

```
> a1 := .4;
  a2 := .15;
> xk := evalm(a1*t1^k*v1+a1*t2^k*v2);
> urban := xk[1,1];
  rural := xk[2,1];
>
```

Maple note: Maple treats a column vector as a matrix with a single column. So the components of a column vector such as the vector *xk* above have two subscripts, where the second subscript is always 1 since a column vector has only one column.

A plot of the urban and rural populations reveals the long-term steady state as the two curves approach the horizontal lines $y = .6$ and $y = .4$ asymptotically.

```
> plot([urban,rural],k=0..100,y=0..1);
>
```

====================

The eigenvector decomposition also shows why the limit of the iterates x_k is $\begin{bmatrix}.6\\.4\end{bmatrix}$ for <u>every</u> initial state x_0. The eigenvector decomposition formula (Theorem 7) for the migration matrix M has the form

$$x_k = a_1\, 1^k \begin{bmatrix}1.5\\1\end{bmatrix} + a_2\, .85^k \begin{bmatrix}-1\\1\end{bmatrix}$$

for every x_0; only the coefficients a_1 and a_2 change when x_0 changes. So, as k becomes large, the iterates x_k get closer and closer to the limit $a_1\begin{bmatrix}1.5\\1\end{bmatrix}$. Furthermore, Since the components of x_0 add to 1 and are non-negative, the same is true for every x_k and hence also for the limit. Therefore, since $1.5\,a_1 + a_1 = 1$, $a_1 = .4$ and the steady-state limit is always $.4\begin{bmatrix}1.5\\1\end{bmatrix}=\begin{bmatrix}.6\\.4\end{bmatrix}$.

To see the long-term behavior of the urban and rural populations under very different initial conditions, execute the following code, which repeats the code in Example 2A. Note expecially the limits, which are again .6 and .4, respectively.

```
> x0 := colvector([.95,.05]);
> a := evalm(P^(-1)&*x0);
> a1 := .4;
```

```
  a2 := -.35;
> xk := evalm(a1*t1^k*v1+a2*t2^k*v2);
> urban := xk[1,1];
  rural := xk[2,1];
> plot([urban,rural],k=0..100,y=0..1);
>
```

Exercise 2.1: (a) For the stochastic matrix N and initial state x_0 below, find the eigenvector decomposition for the iterates $x_k = N^k\, x_0$.

(b) From your answer to (a), find the steady-state limit of the iterates x_k. Observe that this limit is the same for every x_0.

```
> N := matrix([[.9,.4],[.1,.6]]);
> x0 := colvector([.5,.5]);
>
```

Student Workspace

Answer 2.1

Section 3: Discrete Dynamical Systems and Their Trajectories

In this section, we enlarge the scope of the discussion of Section 2 by putting it in a much broader context. Rather than considering only mathematical models of the form $x_{k+1} = A\, x_k$, where A is a stochastic matrix, we allow A to be any square matrix whatsoever. We also introduce the terminology that is normally used in this broader context:

A *discrete dynamical system* is a sequence of vectors x_k related to one another by a square matrix A as follows:

$$x_{k+1} = A\, x_k, \quad \text{for k = 0, 1, 2, ...}$$

If we think of k as representing time in some units, we can think of a dynamical system as describing the evolving behavior over time of the set of variables represented by the vectors x_k. We refer to this behavior as the "dynamics" of the "system" of vectors x_k.

```
>
```

=====================

Example 3A: (a) For the matrix A and initial vector x_0 below, find the eigenvector decomposition of the iterates $x_k = A^k\, x_0$.

(b) From this decomposition, determine the long-term behavior of the points x_k.

```
> A := matrix([[.5250, .1500], [-.1875, .9750]]);
> x0 := colvector([1.5,1]);
```

Solution: (a) First we find the eigenvalues and associated eigenvectors of A:

```
> eigenvals(A);
> Id := identmat(2):
  nullbasis(A-.6*Id);
  nullbasis(A-.9*Id);
```

Therefore we have the eigenvector decomposition $x_k = a_1 \; .6^k \begin{bmatrix} 2 \\ 1 \end{bmatrix} + a_2 \; .9^k \begin{bmatrix} .4 \\ 1 \end{bmatrix}$. Since both $.9^k$ and $.6^k$ have limit 0, the iterates x_k tend toward the zero vector in the limit. To complete part (a), here's the computation of the coefficients a_1 and a_2:

```
> v1 := colvector([2,1]);
  v2 := colvector([.4,1]);
> P := augment(v1,v2);
> a := evalm(P^(-1)&*x0);
>
```

====================

As in Section 2, we could plot the components of x_k as functions of k. However, in this section we will exploit a different kind of picture, one that displays better the "dynamics" of the system. We will plot the "trajectory" of the vectors x_k, which is a plot of the points x_0, x_1, x_2, ... Such a trajectory is much like the plot of a parametrized curve $\mathrm{r}(t) = [\mathrm{x}(t), \mathrm{y}(t)]$ in which we think of t as time; here we can think of k as time.

The command trajectory(A,x0,n) plots the trajectory of the vectors x_k for k from 0 to n as an animation:

```
> trajectory(A,x0,20);
>
```

====================

Example 3B: For the matrix A and initial vector x_0 in Example 3A, use the eigenvector to explain the behavior we saw above of the trajectory of the points x_k.

Solution: Since $x_k = a_1 \; .6^k \begin{bmatrix} 2 \\ 1 \end{bmatrix} + a_2 \; .9^k \begin{bmatrix} .4 \\ 1 \end{bmatrix}$ and since both eigenvalues are less than one in absolute value, the points x_k will converge to the origin for every x_0. Furthermore, since $.6 < .9$, the points x_k will be close to $a_2 \; .9^k \begin{bmatrix} .4 \\ 1 \end{bmatrix}$ for large k, which tells us that x_k approaches the origin along a line with direction vector $\begin{bmatrix} .4 \\ 1 \end{bmatrix}$. The following picture, which includes the two eigenvectors, shows this behavior more clearly:

```
> p1 := trajectory(A,x0,20):
  p2 := drawvec2d(v1,v2,headcolor=blue):
  display([p1,p2]);
>
```

====================

Exercise 3.1: Carry out an analysis in the style of Example 3B for the matrix B below. That is, find the eigenvector decomposition for x_k, and use it to explain the behavior of the trajectory of x_k.

```
> B := matrix([[3.1,-1.0],[5.0,-1.4]]);
```

```
[ >
```

Student Workspace

Answer 3.1

====================

Example 3C: Although the trajectory for the above matrix B is away from the origin, the fact that one of the eigenvalues is between 0 and 1 tells us that, for certain initial vectors x_0, the trajectory can be toward the origin, at least for a while. Below, we choose x_0 very close to the eigenvector $\begin{bmatrix} .4 \\ 1 \end{bmatrix}$, which is associated with the smaller eigenvalue $\lambda = .6$:

```
> x0 := colvector([.41,1]);
> trajectory(B,x0,30);
[ >
```

====================

Exercise 3.2: Carry out an analysis in the style of Example 3B for the matrix C below. That is, find the eigenvector decomposition for x_k, and use it to explain the behavior of the trajectory of x_k.

```
[ > C := matrix([[.9,.2],[.1,.8]]);
[ >
```

Student Workspace

Answer 3.2

The following terminology will be handy for describing the different types of dynamics associated with a 2 by 2 matrix A. We assume here that A has two real eigenvalues, λ_1 and λ_2.

- If both eigenvalues are less than 1 in absolute value, we say that the dynamical system is *attracted* to the origin. (See, for example, matrix A in Example 3A.)
- If both eigenvalues are greater than 1 in absolute value, we say that the dynamical system is *repelled* by the origin.
- If one eigenvalue is greater than 1 in absolute value and the other is less than 1 in absolute value, we say that the dynamical system has a *saddle point* at the origin. (See, for example, matrix B in Exercise 3.1.)

Here's a system whose dynamics are more complicated, although we will still say that the system is attracted to the origin. As we will see in a later module, the spiral behavior is caused by the presence of complex eigenvalues.

```
[ > A := matrix([[.8,.5],[-.1,1.0]]);
[ > trajectory(A,colvector([1,1]),40);
[ >
```

Section 4: Predator-Prey Model : Foxes and Rabbits

In this section we consider a dynamical systems model that simulates the interactions between two populations, a population of foxes (the predators) and a population of rabbits (the prey). We denote the fox and rabbit populations at time k by $x_k = \begin{bmatrix} F_k \\ R_k \end{bmatrix}$, where k is in years and where F_k is the number of foxes after k years and R_k is the number of rabbits (in hundreds) after k years. Also we suppose that:

$$F_{k+1} = .6\, F_k + .4\, R_k$$

$$R_{k+1} = -.125\, F_k + 1.2\, R_k$$

```
>
```

The term .6 F_k in the first equation says that in the absence of rabbits for food only 60% of the foxes would survive each year. On the other hand, the presence of rabbits increases the number of foxes, which is represented by the term .4 R_k. Looking at the second equation, we see the term 1.2 R_k, which represents a 20% growth of rabbits per year if there were no foxes. Finally the effect of the foxes on the rabbits is represented by the term $-.125\, F_k$; thus, the greater the number of foxes, the more the rabbit population will decrease in a given year.

Below is the transition matrix M for the fox and rabbit populations.

```
> M := matrix([[.6,.4],[-.125,1.2]]);
```

Let's find the eigenvectors of the iterates x_k:

```
> eigenvals(M);
> Id := identmat(2):
  nullbasis(M-1.1*Id);
  nullbasis(M-.7*Id);
>
```

Therefore

$$x_k = a_1\, 1.1^k \begin{bmatrix} .8 \\ 1 \end{bmatrix} + a_2\, .7^k \begin{bmatrix} 4 \\ 1 \end{bmatrix}$$

Since one eigenvalue is larger than 1 and the other is smaller than 1, the origin is a saddle point for this dynamical system. More specifically, since x_k is approximately equal to $a_1\, 1.1^k \begin{bmatrix} .8 \\ 1 \end{bmatrix}$ for large values of k, both the fox and rabbit populations grow at a yearly rate of 10% (except in the special case when $a_1 = 0$), and the populations are in the ratio of 8 foxes to every 1000 rabbits. Note, moreover, that these conclusions are independent of the initial population x_0. Let's look at the trajectory in which we start with 50 foxes and 1500 rabbits:

```
> x0 := colvector([50,15]);
> trajectory(M,x0,20);
```

```
>
```

Exercise 4.1: In the above trajectory, observe that the fox population decreases dramatically at first, before both populations eventually get larger and larger. Note the contrasting situation below where we start with 5 foxes and 1300 rabbits. Give a plausible explanation for why these different initial conditions lead to such different trajectories.

```
> trajectory(M,colvector([5,13]),20);
>
```

Student Workspace

Answer 4.1

Problems

Problem 1: Coordinates relative to a basis

Consider the vectors v_1, v_2, v_3 below.

(a) Check that $\boldsymbol{B} = \{ v_1, v_2, v_3 \}$ is a basis for R^3.

(b) Use the approach in Section 1 to find the coordinates of the point [1, 10, 6] relative to the basis ***B***.

(c) By a similar approach, find the standard coordinates of the point whose coordinates relative to the basis ***B*** are [3, −1, 2].

```
> v1 := colvector([1,0,1]);
> v2 := colvector([1,2,0]);
> v3 := colvector([0,1,2]);
>
```

Student Workspace

Problem 2: Eigenvector decomposition for 3 by 3 matrices

(a) Write out the eigenvector decomposition for $x_k = A^k\, x_0$, where A is the 3 by 3 matrix below, and x_0 is the vector below. (Use the eigenvects(..) command.)

```
> A := matrix([[3,2,2],[1,4,1],[ -2,-4,-1]]);
> x0 := colvector([4,2,-5]);
>
```

(b) (By hand) Derive the eigenvector decomposition formula in Theorem 7 for the case of a general 3 by 3 matrix.

Student Workspace

Problem 3: More eigenvector decompositions

For each of the matrices below,

(a) Find the eigenvector decomposition of the iterates $x_k = M^k\, x_0$, where x_0 is the vector whose first component is 1 and the rest are 0.

(b) From this decomposition, determine the long-term behavior of the points x_k. (Do they approach a limit? Do they eventually follow some straight line? Is the system attracted to the origin or repelled by the origin or is the origin a saddle point?)

(c) Determine whether the long-term behavior is independent of the choice of x_0. Justify your conclusion.

```
> A := matrix([[1,-.5],[1,2.5]]);
> B := matrix([[.8,-.2,-.2],[-.2,2,.8],[.2,-1.4,-.2]]);
> C := matrix([[.85,.15,.05],[.05,.75,.05],[.1,.1,.9]]);
> E := matrix([[-1.0,1.2,-2.0],[.8,0,1.0], [1.6,-1.2,2.6]]);
>
```

Student Workspace

Problem 4: More foxes and rabbits

This problem is a follow-up to the fox versus rabbit model presented in Section 4.

(a) Determine the long-term behavior of the model when the [2, 2] entry of the matrix M is changed from 1.2 to 1.05. (Assume that a degraded environment lowers the fecundity of the rabbits.) Base your analysis on an eigenvector decomposition of x_k.

(b) Determine a value of the [2, 2] entry that leads to constant levels of the fox and rabbit populations, so that eventually neither population is changing. What is the ratio of the sizes of the populations in this case?

Student Workspace

Problem 5: Leslie population model

We consider an animal population whose size and age distribution is determined by characteristics of the female. (We assume there will always be enough males to propagate the species.) The maximal lifespan of the Lesser Horned Beetle (LHB, for short) is 4 years. Of the newborn female LHB's, 38% live to one year; of the one-year old females, 30% live to two; of the two-year old females, 25% live to three. Furthermore, on average, each one-year-old female produces 2 female offspring and each two-year-old female produces 2 female offspring.

If $x_k = [a_k, b_k, c_k, d_k]$ denotes the number of female LHB's in each of the four age categories after k years, we have the equations

$$a_{k+1} = 2\, b_k + 2\, c_k$$
$$b_{k+1} = .38\, a_k$$
$$c_{k+1} = .3\, b_k$$
$$d_{k+1} = .25\, c_k$$

(a) From the above system of equations, find the transition matrix M such that $x_{k+1} = M\, x_k$. (M is called a *Leslie* matrix.)

(b) Determine the long-term behavior of x_k. Base your analysis on an eigenvector decomposition of x_k. (Use the eigenvects(..) command.)
(c) Using the initial state vector $x_0 = [4, 3, 2, 1]$, find explicit formulas for a_k, b_k, c_k, d_k and plot them. (**Warning:** You will need to use integerplot(..) instead of plot(..), because some of your terms in a_k, b_k, c_k, d_k will be of the form λ^k, where λ is negative. The plot(..) command will try to use noninteger values of k, which makes λ^k undefined.)
Student Workspace

Pencil and Paper Practice Questions

Note: Maple is not required for any of these problems. Do all of these by hand.

1. In each of the parts (i), (ii), and (ii) below, we list the output of the eigenvects(..) command for a square matrix A. For each of these,
(a) find the eigenvector decomposition for $x_k = A^k x_0$; and
(b) describe the long-term behavior of the iterates x_k, including a description of the trajectory in as much detail as is possible when you do not know the initial vector x_0.

(i) $\left[-\frac{1}{4}, 1, \{[-4, 1]\}\right], \left[\frac{1}{2}, 1, \{[-1, 1]\}\right]$

(ii) $[-1, 1, \{[1, -4]\}], \left[\frac{5}{4}, 1, \{[2, 1]\}\right]$

(iii) $\left[\frac{1}{2}, 2, \{[2, 0, 1], [-1, 1, 0]\}\right], \left[\frac{3}{2}, 1, \{[1, 0, 1]\}\right]$

Linear Algebra Modules Project
Chapter 6, Module 3

Diagonalization and Similarity

Purpose of this module

The purpose of this module is to study the process by which the eigenvectors of a matrix can be used to diagonalize a matrix. We then study the more general concept of similar matrices.

Prerequisites

Basic properties of eigenvalues and eigenvectors.

Commands used in this module

```
> restart: with(linalg): with(lamp): with(plots):
>
```

Tutorial

Section 1: Diagonalizing a Matrix

We have seen that the eigenvalues and eigenvectors of a matrix provide us with valuable insights into how matrix acts geometrically as a transformation of vectors. In this section we study how a matrix can be "diagonalized" and then factored into a product of matrices that reveals all the eigenvector and eigenvalue information about the matrix in a compact and useful form.

===================

Example 1A (How to diagonalize a matrix): For the matrix A defined below, we first calculate its eigenvalues and eigenvectors:

```
> A := matrix([[1,4],[2,3]]);
> eigenvects(A);
>
```

So A has eigenvalue $\lambda_1 = -1$ with eigenvector $v_1 = \begin{bmatrix} -2 \\ 1 \end{bmatrix}$ and eigenvalue $\lambda_2 = 5$ with eigenvector $v_2 = \begin{bmatrix} 1 \\ 1 \end{bmatrix}$.

```
> v1 := colvector([-2,1]);
  v2 := colvector([1,1]);
```

Next we construct the matrix P whose columns are the eigenvectors v_1 and v_2.

```
> P := augment(v1,v2);
```

Since the columns of P are linearly independent, P has an inverse. We then calculate the product $P^{(-1)} A\ P$:

```
[ > K := evalm(P^(-1)&*A&*P);
[ >
```

K is the diagonal matrix we are looking for. Notice that K is not only diagonal but the diagonal entries are in fact the eigenvalues of A, appearing in the same order as their corresponding eigenvectors appear as columns in P.

==================

The following theorem summarizes the method of Example 1A for diagonalizing a square matrix:

> ***Theorem 8***: If A is an n by n matrix with n linearly independent eigenvectors, then A can be diagonalized. Specifically, if P is a matrix whose columns are n linearly independent eigenvectors of A, then $P^{(-1)} A\ P = K$, where K is the diagonal matrix whose diagonal entries are the corresponding eigenvalues.

==================

Example 1B: Diagonalize the matrix A below by the method of Example 1A and Theorem 8.

```
[ > A := matrix([[1,-2,4,2],[-2,1,4,2],[0,-2,5,2],[-2,-2,4,5]]);
[ >
```

Solution: The eigenvects(..) command gives us four linearly independent eigenvectors (bases for the three eigenspaces).

```
[ > eigenvects(A);
[ > v1 := colvector([1,0,1,-1]);
    v2 := colvector([0,1,0,1]);
  > v3 := colvector([1,1,1,1]);
    v4 := colvector([1,1,0,1]);
[ >
```

Maple Note: The eigenvects(..) command doesn't always produce the same output. If your result does not coincide with the vectors we defined above, you might try executing eigenvects(A) again until you get the vectors v_1, v_2, v_3, v_4 above.

Next we construct the matrix P and check that the columns of P are linearly independent:

```
[ > P := augment(v1,v2,v3,v4);
[ > matsolve(P,colvector([0,0,0,0]));
[ >
```

Since $P\ x = 0$ has only the zero solution, the columns of P are linearly independent and hence P is invertible. Next we compute the diagonal matrix $K = P^{(-1)} A\ P$. Note that the diagonal entries of K are indeed the eigenvalues produced by the eigenvects(..) command and that they appear in K in the same order as their corresponding eigenvectors in P.

```
[ > K := evalm(P^(-1)&*A&*P);
[ >
```

==================

Exercise 1.1: The matrix Q below has the columns of P above but in a different order. Predict what the diagonal matrix $Q^{(-1)} A\, Q$ will be before doing any computation.

```
> Q := augment(v2,v3,v4,v1);
>
```

Student Workspace

Answer 1.1

Exercise 1.2: Below is a four-step proof of Theorem 8 for 2 by 2 matrices. Give a brief justification for each step. We are given a 2 by 2 matrix A with linearly independent eigenvectors v_1, v_2 and corresponding eigenvalues λ_1, λ_2. Let $P = [v_1, v_2]$ and $K = \begin{bmatrix} \lambda_1 & 0 \\ 0 & \lambda_2 \end{bmatrix}$. Then

(1) $A\,P = A\,[v_1, v_2] = [A\,v_1, A\,v_2]$

(2) $P\,K = [v_1, v_2] \begin{bmatrix} \lambda_1 & 0 \\ 0 & \lambda_2 \end{bmatrix} = [\lambda_1\, v_1, \lambda_2\, v_2]$

(3) $A\,P = P\,K$

(4) $P^{(-1)} A\,P = K$

```
>
```

Student Workspace

Answer 1.2

==================

Example 1C: The matrix A below reflects vectors in R^2 across the line through the origin making an angle of 60 degrees with the x axis.

```
> A := reflectmat(Pi/3);
> eigenvects(A);
```

As before, we use the eigenvectors of A to construct the matrix P and carry out the diagonalization:

```
> v1 := colvector([1,sqrt(3)]);
> v2 := colvector([-sqrt(3),1]);
> P := augment(v1,v2);
> K := evalm(P^(-1)&*A&*P);
>
```

The resulting diagonal matrix is not a surprise, since the eigenvalues of a reflection matrix are simply 1 and -1, corresponding to eigenvectors parallel and perpendicular to the line of reflection.

==================

Exercise 1.3:

(a) The diagonal matrix K that we found in Example 1C is also a reflection matrix. What is its line of reflection?

(b) Diagonalize the two reflection matrices given below.

```
> A := reflectmat(Pi/4);
```

```
> B := reflectmat(Pi/2);
>
```

(c) As the above examples suggest, all reflection matrices have the same diagonalized form, with 1 and -1 along the diagonal. To demonstrate this, diagonalize the general reflection matrix R below, using the eigenvectors v_1 and v_2 below to construct P. To simplify the final result, use the command map(simplify,K).

```
> R := reflectmat(theta);
> v1 := colvector([cos(theta),sin(theta)]);
  v2 := colvector([-sin(theta),cos(theta)]);
>
```

Student Workspace

Answer 1.3

Diagonalizing the Powers of a Matrix

Suppose we have diagonalized a matrix A in the usual way:

$$P^{(-1)} A\, P = K$$

If we multiply this equation on the left by P and on the right by $P^{(-1)}$, we get a factorization of A into a product of three matrices:

$$A = P\, K\, P^{(-1)}$$

From this factorization of A, we can derive a similar factorization of A^2:

$$A^2 = A\, A = P\, K\, P^{(-1)}\, P\, K\, P^{(-1)} = P\, K\, I\, K\, P^{(-1)} = P\, K^2\, P^{(-1)}$$

```
>
```

It is not surprising that A and A^2 have the same diagonalizing matrix P, since A and A^2 have the same eigenvectors (Problem 7 in Module 1). Similarly, for any positive integer power n, we have a factorization of A^n:

- $A^n = P\, K^n\, P^{(-1)}$

Exercise 1.4: For the matrix A in Example 1A (repeated below), use the formula $A^n = P\, K^n\, P^{(-1)}$ to compute A^5.

```
> A := matrix([[1,4],[2,3]]);
>
```

Student Workspace

Answer 1.4

In the next example we use the above formula for A^n to study the powers of a Markov matrix.

==================

Example 1D: Find a formula for the *n*th power of the Markov matrix *M* below. Also find the limit of M^n as *n* goes to infinity.

```
> M := matrix([[94/100,9/100],[6/100,91/100]]);
> eigenvects(M);
>
```

So *M* has eigenvalues 0.85 (i.e., 17/20) and 1 with eigenvectors $\begin{bmatrix} 1 \\ -1 \end{bmatrix}$ and $\begin{bmatrix} 1.5 \\ 1 \end{bmatrix}$, respectively. Since $M\,x = x$ for the eigenvector $x = \begin{bmatrix} 1.5 \\ 1 \end{bmatrix}$, this eigenvector is a steady-state vector for the Markov matrix *M*. More properly, if we multiply this vector by 0.4 so its entries add to 1, we see that $\begin{bmatrix} .6 \\ .4 \end{bmatrix}$ is the steady-state vector. However, we want to see how this steady-state vector arises directly from the powers of *M*, and so we diagonalize *M*:

```
> P := matrix([[1,1.5],[-1,1]]);
> K := evalm(P^(-1)&*M&*P);
>
```

Since the *n*th power of a diagonal matrix is found by simply taking the *n*th powers of the diagonal entries, we have derived the following formula for M^n:

$$M^n = P\,K^n\,P^{(-1)} = P\begin{bmatrix} .85^n & 0 \\ 0 & 1^n \end{bmatrix} P^{(-1)}$$

Since the limit of K^n, as *n* goes to infinity, is $\begin{bmatrix} 0 & 0 \\ 0 & 1 \end{bmatrix}$, the limit of M^n is $P\begin{bmatrix} 0 & 0 \\ 0 & 1 \end{bmatrix}P^{(-1)}$. Thus:

```
> L := matrix([[0,0],[0,1]]);
> evalm(P&*L&*P^(-1));
>
```

In particular, note that the columns of this limit matrix are identical and that they are the steady state vector $\begin{bmatrix} .6 \\ .4 \end{bmatrix}$. This shows why the limit of the state vectors $M^k\,x_0$ is $\begin{bmatrix} .6 \\ .4 \end{bmatrix}$ for every initial state x_0 (see Problem 1).

==================

```
>
```

Section 2: Similar Matrices

In Section 1 we found that we can diagonalize an *n* by *n* matrix, provided we can find *n* linearly independent eigenvectors for the matrix. If the matrix does not have *n* linearly independent eigenvectors, we cannot diagonalize it. However, even in this case we can carry out a process like diagonalization which has useful consequences.

If we have a matrix A and any invertible matrix P, we can calculate the product $P^{(-1)} A P$. The resulting matrix may not be a diagonal matrix, but it does have an interesting relationship to A.

> ***Definition:*** Two n by n matrices A and B are ***similar*** if there is an invertible matrix P such that:
>
> $$P^{(-1)} A P = B, \quad \text{or equivalently,} \quad A = P B P^{(-1)}$$

In particular, when we diagonalize a matrix A by finding a matrix P such that $P^{(-1)} A P = K$ is diagonal, we will now say that A is *similar* to the diagonal matrix K. Let's revisit Example 1C in Section 1:

```
> 
```

====================

Example 2A: In Example 1C, we diagonalized the reflection matrix A (repeated below) that reflects vectors across the line making an angle of 60 degrees with the x axis. We used the diagonalizing matrix $P = [v_1, v_2]$, where v_1, v_2 (repeated below) are linearly independent eigenvectors of A:

```
> A := reflectmat(Pi/3);
> v1 := colvector([1,sqrt(3)]);
  v2 := colvector([-sqrt(3),1]);
> P := augment(v1,v2);
> evalm(P^(-1)&*A&*P);
> 
```

So the reflection matrix A is similar (according to the above definition) to the matrix that reflects vectors across the x axis. If we change P slightly, however, we can interpret the product $P^{(-1)} A P$ geometrically and see more clearly why this product is the matrix that reflects across the x axis.

```
> Q := rotatemat(Pi/3);
> K := evalm(Q^(-1)&*A&*Q);
> 
```

Note that the columns of the rotation matrix Q are just the columns of P multiplied by 1/2; so they are again eigenvectors of A. Moreover, we can more readily interpret the product $Q^{(-1)} A Q$ geometrically. Let v_1 be some vector in the plane, and let $v_2 = Q v_1$, and $v_3 = A v_2 = A Q v_1$, and $v_4 = Q^{(-1)} v_3 = Q^{(-1)} A Q v_1$. We then have the following picture:

```
> v1 := colvector([3,1]):
  v2 := evalm(Q&*v1):
  v3 := evalm(A&*v2):
  v4 := evalm(Q^(-1)&*v3):
> p1 := drawvec2d([v1,label="v1"],[v2,label="v2"],
    [v3,label="v3"], [v4,label="v4"]):
> p2 := plot(tan(Pi/3)*t,t=-1..2,color=red):
```

```
> display({p1,p2},view=[-1..3.5,-2..4]);
>
```

Thus, the equation $K\, v_1 = Q^{(-1)}\, A\, Q\, v_1$ says that if we rotate a vector v_1 counter-clockwise by 60 degrees to v_2, then reflect v_2 across the 60-degree line to v_3, and then rotate v_3 back by 60 degrees to v_4, the overall result will be a reflection across the x axis, which is exactly what $K\, v_1$ is.

====================

Of course there is nothing special about the reflection across the angle of 60 degrees. In Exercise 1.3(c) we showed that every reflection matrix is similar to the special reflection matrix about the x axis, and therefore all reflection matrices are similar to one another (see Problem 5). This suggests that similar matrices are much alike. As we will see, when two matrices that are similar, in the sense of the above definition, they have many of the same properties. The following theorem describes one such property:

Theorem 9: If A and B are similar matrices, they have the same characteristic polynomial and hence the same eigenvalues with the same algebraic multiplicities.

```
>
```

The proof, which depends only on the property of determinants that $\det(A\,B) = \det(A)\det(B)$, is:

$$\begin{aligned}\det(B - \lambda\, I) &= \det(P^{(-1)}\, A\, P - \lambda\, P^{(-1)}\, P)\\ &= \det(P^{(-1)}\,(A - \lambda\, I)\, P)\\ &= \det(P^{(-1)})\det(A - \lambda\, I)\det(P)\\ &= \det(A - \lambda\, I)\end{aligned}$$

====================

Example 2B: Given any square matrix A, we can construct a similar matrix B by choosing any invertible n by n matrix for P, and defining B to be $P^{(-1)}\, A\, P$. Below we construct B in this way and check that A and B have the same characteristic polynomial:

```
> A := matrix([[3,-4],[1,-2]]);
> P := matrix([[2,3],[1,2]]);
```

Note that P is invertible since its columns are linearly independent.

```
> B := evalm(P^(-1)&*A&*P);
> charpoly(A,t);
> charpoly(B,t);
>
```

====================

Do similar matrices also have the same eigenvectors? Execute the next input region and take a close look at the result. Notice that while the eigenvalues and their algebraic multiplicities are the same, their eigenvectors are quite different.

```
> eigenvects(A);
> eigenvects(B);
>
```

Exercise 2.1: (a) Check that when each of the above eigenvectors of A is multiplied by $P^{(-1)}$, the result is an eigenvector of B with the same eigenvalue.

(b) (By hand) Show in general that if x is an eigenvector of a matrix A with eigenvalue λ, then the vector $P^{(-1)}x$ is an eigenvector of $P^{(-1)} A P$ with the same eigenvalue.

Student Workspace

Answer 2.1

We can apply Exercise 2.1 as follows. The matrix function $P^{(-1)}$ sends each eigenspace of the matrix A onto an eigenspace of the matrix $B = P^{(-1)} A P$ (for the same eigenvalue). Since $P^{(-1)}$ is an invertible function, it will also send a basis of the eigenspace of A onto a basis of the eigenspace of B. This proves:

> ***Theorem 10:*** If A and B are similar matrices, their eigenvalues have the same geometric multiplicities.

Exercise 2.2: The matrices A and B below are similar, and $\lambda = 3$ is an eigenvalue for both. Check that $\lambda = 3$ has the same geometric multiplicity for A and B.

```
> A := matrix([[5,-4,2],[4,-5,4],[2,-4,5]]);
> P := matrix([[-1,0,0],[-1,0,-1],[1,1,-1]]);
> B := evalm(P^(-1)&*A&*P);
>
```

Student Workspace

Answer 2.2

Section 3: Geometric Meaning of Similarity and Diagonalization

In this section we will think of matrices as matrix transformations. That is, if A is an n by n matrix, we will study the behavior of the function T defined by

$$\mathrm{T}(x) = A\,x, \text{ where } x \text{ ranges over all vectors in } R^n.$$

We will see that by diagonalizing the matrix A we can understand the behavior of the matrix transformation T geometrically. To see how to interpret a matrix transformation geometrically, we first consider diagonal matrix transformations.

====================

Example 3A: Describe the geometric behavior of the diagonal matrix K below.

```
> K := diag(2,-1);
```

Solution: As a function, K takes vectors on the x axis to twice themselves and reverses the direction of vectors on the y axis:

```
> evalm(K&*colvector([x,0]));
  evalm(K&*colvector([0,y]));
>
```

So the behavior of *K* as a function is easy to describe when *K* is applied to vectors along either axis. (Notice, of course, that the *x* axis is an eigenspace of *K* with associated eigenvalue 2, and the *y* axis is an eigenspace of *K* with eigenvalue -1.) Furthermore, once we know how *K* behaves when applied to vectors on either axis, we can also describe how *K* behaves when applied to other vectors, since all vectors in R^2 can be decomposed into their *x* and *y* components. For example, if *K* is applied to the vector $\begin{bmatrix} -1 \\ 3 \end{bmatrix}$, *K* doubles the *x* component to -2 and reverses the sign of the *y* component to -3:

```
> evalm(K&*colvector([-1,3]));
>
```

More geometrically, we can say that *K* stretches figures by a factor of 2 in the *x* direction and reflects figures across the *x* axis (i.e., in the direction of the *y* axis). For example, here's a picture of *K* applied to the house figure:

```
> transform(K,house);
>
```

====================

Exercise 3.1: Describe the behavior of the diagonal matrix transformation *L* below. Specifically:
(a) What are the eigenspaces of *L*, and what does *L* do to vectors in each eigenspace?

(b) Use your answer in (a) to describe what *L* does to the vector $\begin{bmatrix} 3 \\ 4 \\ 5 \end{bmatrix}$.

```
> L := diag(-2,3,3);
>
```

Student Workspace

Answer 3.1

Now suppose *A* is a matrix that can be diagonalized. We will see that the behavior of such a matrix transformation *A* can also be described geometrically. Consider the following example.

```
> A := matrix([[1,-2],[-1,0]]);
> eigenvects(A);
```

So the eigenvalues of *A* are 2 and -1 with associated eigenvectors *u* and *v* below:

```
> u := colvector([-2,1]);
  v := colvector([1,1]);
> P := augment(u,v);
> K := evalm(P^(-1)&*A&*P);
>
```

The columns of the matrix *P* are linearly independent eigenvectors of *A*, and *K* is the diagonal matrix similar to *A* that *P* produces. We saw in Example 3A how to describe the behavior of *K*, and now we want to use that knowledge to describe the behavior of *A*. Since *u* is an eigenvector of *A* with eigenvalue 2, *A* takes vectors in the eigenspace Span$\{u\}$ to twice themselves. And, since *v* is an eigenvector of *A* with eigenvalue -1, *A* reverses the direction of vectors in the eigenspace Span$\{v\}$.

Furthermore, we can describe what A does to every vector in R^2 by decomposing vectors into their u and v components.

====================

Example 3B: Describe what A does to the vector $y = \begin{bmatrix} 5 \\ 2 \end{bmatrix}$ by decomposing y into its u and v components, where u and v are the above eigenvectors of A.

Solution: To solve $a\,u + b\,v = y$ for a and b, we write the equation in the form $[u, v]\begin{bmatrix} a \\ b \end{bmatrix} = y$ and solve:

```
[ > y := colvector([5,2]);
[ > x := evalm(P^(-1)*y);
```

Thus $a = -1$ and $b = 3$. So A doubles the u component of y from -1 to -2 and reverses the sign of the v component of y from 3 to -3. Let's check this result:

```
[ > evalm(-2*u-3*v);
[   evalm(A&*y);
```

Since $-2\,u - 3\,v = [u, v]\begin{bmatrix} -2 \\ -3 \end{bmatrix} = P\,K\,x$, we can also compute $A\,y$ as follows:

```
[ > evalm(P&*K&*x);
[ >
```

In a sense, therefore, A behaves very much like the similar diagonal matrix K, and its behavior is almost as simple to describe. The diagonal matrix K behaves especially simply when applied to vectors on the x and y axes, while A behaves especially simply when applied to vectors on the u and v axes.

====================

Let's compare the behavior of K and A visually by applying both to the "bug" figure:

```
[ > transform(K,bug);
[ > transform(A,bug);
[ >
```

Although the results are not identical, you can see the similarities. Both matrices reflect the bug in one direction and stretch it in another direction. Also, both change the area of the bug by the same factor, since $\det(P^{(-1)}\,A\,P) = \det(A)$ and the magnitude of the determinant equals the ratio of the image area to the original area.

We can even argue that, in a sense, A and K represent the <u>same</u> function in two different coordinate systems, the standard system and the uv system. To see this, let's look at a smaller version of the house figure together with the standard basis vectors, e_1 and e_2:

```
[ > shed := .4*house;
[ > p1 := drawmatrix(shed,figcolor=black):
  > p2 := drawvec2d([1,0],[0,1],headcolor=black):
[ > display([p1,p2],view=[-3..2,0..3]);
[ >
```

Next, we look at the shed figure in *uv* coordinates:

```
[ > uv_shed := evalm(P&*shed);
```

So the coordinates of *uv*_shed relative to the basis { u, v } are the same as the coordinates of the shed figure relative to the basis { e_1, e_2 }. Then we draw *uv*_shed together with the basis vectors u and v, and we also draw shed and *uv*_shed together:

```
> p3 := drawmatrix(uv_shed):
> p4 := drawvec2d([u,label="u"],[v,label="v"]):
> display([p3,p4]);
[ > display([p1,p2,p3,p4]);
[ >
```

Finally, we apply the original matrix A to *uv*_shed, and observe that A behaves just as K did; that is, A reflects across the u axis in the direct of the v axis and stretches by a factor of 2 in the direction of the u axis:

```
> p5 := transform(A,uv_shed):
> display([p3,p4,p5]);
[ >
```

Exercise 3.2:

(a) Use the diagonalization process to show that the matrices R and K below are similar.

```
> R := reflectmat(Pi/4);
> K := reflectmat(0);
[ >
```

(b) Use geometry to explain how R and K can be thought of as essentially the same matrix transformation.

Student Workspace

Answer 3.2

Problems

Problem 1: Markov matrices, powers, and stable vectors

Let M be the Markov matrix below.

(a) Use a diagonalization of M to calculate M^5 and M^{10}.

(b) (By hand) Write down a formula for M^n. Then find the exact limit of M^n as n goes to infinity.

(c) (By hand) Use your result in (b) to explain the following. If x_0 is any vector in R^2 whose components add to 1, then the limit of $M^k x_0$ is always $\begin{bmatrix} \frac{3}{7} \\ \frac{4}{7} \end{bmatrix}$.

```
[ > M := matrix([[1/5,3/5],[4/5,2/5]]);
```

[>

Student Workspace

Problem 2: Square roots of matrices

Recall that a positive real number r has two square roots -- one positive and one negative. For example, both 2 and -2 are square roots of 4. Matrices typically have more square roots, because we have more places to choose signs plus or minus.

(a) (By hand) The 2 by 2 diagonal matrix diag(1,4) has four diagonal square roots (i.e., diagonal matrices whose square is the given matrix). Find them. (Hint: One of the diagonal square roots is diag(-1,2).) For each of these square roots, state its eigenvalues.

(b) (By hand) Write down the eight diagonal square roots of diag(1,4,9).

(c) Find a square root of the matrix A in the input region below. That is find a matrix C such that $C^2 = A$ (Hint: Diagonalize A and find a square root of the diagonal matrix K. Use the square root of K to find C.) Describe how you would find three other square roots of A (but don't bother doing the computation).

```
[ > A := matrix([[13,18],[-6,-8]]);
[ >
```

(d) Apply the idea in (c) to find a square root of the projection matrix projectmat(π/4). By considering the nature of the eigenvalues of this matrix, explain why you get fewer square roots this time.

(e) By the same reasoning as in (a), the 2 by 2 identity matrix I_2 has four diagonal square roots. But the fact that the eigenvalue 1 for I_2 has geometric multiplicity 2 turns out to mean that I_2 has infinitely many choices of square root, and these are related in interesting ways by similarity. Choose any diagonal square root T of I_2, such as diag(-1,1), and any nonsingular matrix S, such as $\begin{bmatrix} 1 & 2 \\ 3 & 4 \end{bmatrix}$. Check that $S^{(-1)}\, T\, S$ is a square root of I_2. Find at least two nondiagonal square roots of I_2 by this method.

(f) (By hand) Prove by matrix algebra that every matrix of the form $S^{(-1)}\, T\, S$, where $T^2 = I_n$ and S is nonsingular, is a square root of I_n (I_n is the n by n identity matrix).

Student Workspace

Problem 3: Construct a matrix from its eigenvalues and eigenvectors

(a) Construct a 2 by 2 matrix A that has the eigenvector [1, 2] with eigenvalue 1 and eigenvector [2, 3] with eigenvalue 4. Hint: Use $P^{(-1)}\, A\, P = K$, where K is diagonal.

(b) (By hand) Suppose you are given n linearly independent vectors $v_1, ..., v_n$ and n scalars $t_1, ..., t_n$. Describe in general how to construct an n by n matrix A with eigenvectors $v_1, ..., v_n$ and corresponding eigenvalues $t_1, ..., t_n$.

Student Workspace

Problem 4: Linear independence of eigenvectors

Suppose v_1 and v_2 are eigenvectors of a square matrix A and that λ_1 and λ_2 are the associated eigenvalues. If $\lambda_1 \neq \lambda_2$, prove that $\{v_1, v_2\}$ is linearly independent. Hint: If $\{v_1, v_2\}$ were linearly dependent, one of the vectors would be a multiple of the other. Show that this implies that both vectors have the same eigenvalue.

Note: This result is a special case of the following more general theorem, which is proved in a similar way:

> ***Theorem 11:*** Suppose A is a square matrix and $\{v_1, ..., v_m\}$ is a set of eigenvectors of A constructed as follows: Select a basis for each eigenspace of A and take all these basis vectors collectively. Then $\{v_1, ..., v_m\}$ is linearly independent.

Student Workspace

Problem 5: Similar to similar implies similar

(By hand) If A, B, C are n by n matrices such that A and B are similar and B and C are similar, prove that A and C are similar. **Caution:** The invertible matrix that relates A and B is not necessarily the same matrix that relates B and C.

Student Workspace

Problem 6: Geometric behavior of a matrix transformation

(a) Describe the geometric behavior of the diagonal matrix transformation K below.
(b) Describe the geometric behavior of the matrix transformation A below. (Imitate the discussion in Section 3.)
(c) For the vector $y = [2, 2, -1]$, find the coordinates of y relative to the eigenvector basis you found in (b). Use these coordinates to describe what A does to the vector y.

```
> K := diag(0,-2,-2);
> A := matrix([[-3,1,-2],[1,-3,2],[2,-2,2]]);
>
```

Student Workspace

Problem 7: Projection matrices

Here is the general 2 by 2 projection matrix, which projects vectors onto the line making an angle of θ radians with the x axis.

```
> A := projectmat(theta);
>
```

(a) Use geometry to explain why two of the eigenvectors for A are $u = \begin{bmatrix} \cos(\theta) \\ \sin(\theta) \end{bmatrix}$ and $v = \begin{bmatrix} -\sin(\theta) \\ \cos(\theta) \end{bmatrix}$.

What are the associated eigenvalues for u and v?

(b) Construct the matrix $P = [u, v]$ to diagonalize A. What familiar matrix transformation is P?
(c) Use the general form of A (i.e., projectmat(theta)) and P to find the diagonal matrix K. In order to get Maple to simplify the result, use map(simplify, K) .
(d) What familiar geometric transformation is K?
(e) It is claimed that all projection matrices are similar. Do you agree? Use your results from parts (a)-(d) to support your answer.
(f) Give a geometric explanation for why the matrix product $P\,K\,P^{(-1)}$ has the same effect as A. Hint: Take an arbitrary vector v_1 and apply $P^{(-1)}$ to v_1. What does $P^{(-1)}$ do to v_1 geometrically? Then apply K. Then apply P.

Student Workspace

Problem 8: Shear matrices

Recall that horizontal shear matrices have the form $\begin{bmatrix} 1 & k \\ 0 & 1 \end{bmatrix}$ where $k \neq 0$.

(a) Find the eigenvalues of $H = \begin{bmatrix} 1 & 1 \\ 0 & 1 \end{bmatrix}$, and find their algebraic and geometric multiplicities. Explain why this matrix cannot be diagonalized.

(b) Show that every horizontal shear matrix, $S = \begin{bmatrix} 1 & k \\ 0 & 1 \end{bmatrix}$ where $k \neq 0$, is similar to the special shear matrix H in (a). (Hint: Use $P = \begin{bmatrix} 1 & 0 \\ 0 & \frac{1}{k} \end{bmatrix}$ as the diagonalizing matrix.) Explain geometrically why $P^{(-1)}\,S\,P = H$, especially describing the effects of P and $P^{(-1)}$.

(c) Show that the vertical shear matrix $V = \begin{bmatrix} 1 & 0 \\ 1 & 1 \end{bmatrix}$ is similar to the horizontal shear matrix H in (a). (Hint: To change the coordinates, use a certain reflection matrix for P .)

(d) Find the eigenvalues of $A = \begin{bmatrix} 3 & -1 \\ 4 & -1 \end{bmatrix}$, and find their algebraic and geometric multiplicities.

(e) Show that the matrix A in (d) is similar to the horizontal shear matrix H in (a). Hint: The equations

$$A\begin{bmatrix} 1 \\ 2 \end{bmatrix} = 1\begin{bmatrix} 1 \\ 2 \end{bmatrix} + 0\begin{bmatrix} 1 \\ 1 \end{bmatrix}, \quad A\begin{bmatrix} 1 \\ 1 \end{bmatrix} = 1\begin{bmatrix} 1 \\ 2 \end{bmatrix} + 1\begin{bmatrix} 1 \\ 1 \end{bmatrix}$$

show that the behavior of A on the two vectors $\begin{bmatrix} 1 \\ 2 \end{bmatrix}$ and $\begin{bmatrix} 1 \\ 1 \end{bmatrix}$ is similar to the behavior of H on the two standard basis vectors $\begin{bmatrix} 1 \\ 0 \end{bmatrix}$ and $\begin{bmatrix} 0 \\ 1 \end{bmatrix}$.

Student Workspace

Pencil and Paper Practice Questions

Note: Maple is not required for any of these problems. Do all of these by hand.

1. When the command eigenvects(A) is applied to the matrix

$$A = \begin{bmatrix} 4 & -1 & 6 \\ 2 & 1 & 6 \\ 2 & -1 & 8 \end{bmatrix}$$

the result is $[2, 2, \{[0, 6, 1], [1, 2, 0]\}]_{9,\,1,\,\{[1,\,1,\,1]\}}$.

(a) Find an invertible matrix P and a diagonal matrix K such that $P^{(-1)} A\, P = K$.

(b) Use your answer to (a) to write a formula for A^n that uses P and K. (Do not simplify your formula.)

2. Justify the following statement algebraically and explain it geometrically: Every 2 by 2 reflection matrix is similar to the matrix $\begin{bmatrix} 1 & 0 \\ 0 & -1 \end{bmatrix}$.

3. The vectors $\begin{bmatrix} 1 \\ 0 \\ -1 \end{bmatrix}, \begin{bmatrix} 0 \\ 1 \\ 2 \end{bmatrix}, \begin{bmatrix} -2 \\ 1 \\ 0 \end{bmatrix}$ are eigenvectors of a 3 by 3 matrix A with associated eigenvalues 2, 3, and -4, respectively.

(a) Find the coordinates of the vector $v = \begin{bmatrix} 1 \\ 0 \\ 3 \end{bmatrix}$ relative to the basis of R^3 consisting of the three eigenvectors above.

(b) Use your answer to (a) to find $A\, v$.

Linear Algebra Modules Project
Chapter 7, Module 1

Orthogonal Vectors and Orthogonal Projections

Purpose of this module

The purpose of this module is to introduce the following concepts and their basic properties: dot product, orthogonal vectors, orthogonal projection, magnitude, and distance in R^n. We use projections to solve a shortest distance problem that will be a key to an important area of applications in a later module.

Prerequisites

Vector algebra, dot product of vectors in R^2 and R^3.

Commands used in this module

```
> restart: with(linalg): with(lamp):
>
```

Tutorial

Section 1: The Dot Product and Orthogonal Vectors in R^n

The definition of the dot product of two vectors in R^n is the natural generalization of the dot product for vectors in R^2 and R^3:

Definition: If $u = [u_1, ..., u_n]$ and $v = [v_1, ..., v_n]$ are vectors in R^n, then the ***dot product*** of u and v, denoted $u.v$, is the sum of the products of the corresponding components of u and v:

$$u.v = u_1\, v_1 + \ldots + u_n\, v_n$$

```
>
```

Note: We should have written the vectors u and v above as column vectors. However, in the text regions of this module, we will write vectors as row vectors simply to save space.

====================

Example 1A : Let $u = [2, 1, 4, -3]$ and $v = [3, 2, 1, 3]$. Then

$$u.v = (2)(3) + (1)(2) + (4)(1) + (-3)(3) = 6 + 2 + 4 - 9 = 3$$

Using Maple:

```
> u := colvector([2,1,4,-3]);
  v := colvector([3,2,1,3]);
> dotproduct(u,v);
>
```

====================

Recall that in R^2 and R^3, two vectors are orthogonal (i.e., perpendicular) if and only if the dot product of the vectors equals 0. We define orthogonality in R^n similarly:

Definition: Vectors u and v in R^n are ***orthogonal*** if $u.v = 0$.

Note that the vectors u and v in Example 1A are not orthogonal since their dot product is 3, not 0.

Exercise 1.1: Change the last component of v (and no other components of v or u) so that $u.v = 0$.

Student Workspace

Answer 1.1

The dot product of two vectors is closely related to matrix multiplication. If we think of u and v as n by 1 matrices, then $u.v$ equals the single entry in the 1 by 1 matrix product $u^T v$:

$$[2 \quad 1 \quad 4 \quad -3]\begin{bmatrix} 3 \\ 2 \\ 1 \\ 3 \end{bmatrix} = [3]$$

```
>
```

Length and Distance in R^n

Recall that for a vector $x = [a, b]$ in R^2 we calculate the length of x, denoted $||x||$, by using the Pythagorean Theorem:

$$||x|| = \sqrt{a^2 + b^2}$$

Alternatively, $||x||$ can be expressed in terms of a dot product, $||x|| = \sqrt{x.x}$, since $x.x = a^2 + b^2$. We define the length of a vector in R^n similarly:

Defintion: The ***length*** (or ***magnitude)*** of a vector $u = [u_1, ..., u_n]$ in R^n is defined to be the square root of $u.u$:

$$||u|| = \sqrt{u.u}$$

[>

Note that if we square both sides of the above identity, we get $\|u\|^2 = u.u$, a useful fact to remember.

The distance between any two vectors in R^n is defined as follows:

Definition: If u and v are vectors in R^n then the ***distance between u and v*** is the magnitude of the vector $u - v$, i.e., $\|u - v\|$.

[>

Exercise 1.2: (By hand) Draw the vectors $u = [1, -3]$ and $v = [6, 4]$. Label the points P and Q corresponding to the heads of the vectors u and v. Now add the vector $u - v$ to your sketch, placing the tail of $u - v$ at the head of the vector v. Calculate the distance from P to Q in the following ways: (a) find the length of the vector $u - v$ using the dot product; (b) use the familiar distance formula $d = \sqrt{(x_1 - x_2)^2 + (y_1 - y_2)^2}$; and (c) use the command mag(u-v).

Student Workspace

Answer 1.2

Rules of Algebra for the Dot Product

The dot product and length satisfy the following rules of algebra (where u, v, w denote vectors and c denotes a scalar):

$$u.v = v.u$$
$$u.(v + w) = u.v + u.w$$
$$(c\,u).v = u.(c\,v) = c\,(u.v)$$
$$\|c\,u\| = |c|\,\|u\|$$

In fact, the second and third of these rules of algebra follow from the corresponding rules for matrix algebra, since the dot product can be treated as a matrix product.

[>

The fundamental theorem relating orthogonality and length in R^2 is the Pythagorean Theorem. This theorem, in vector form, is also true in R^n:

Theorem 1: (The Pythagorean Theorem for R^n) If u and v are orthogonal vectors in R^n, then the square of $\|u + v\|$ equals the sum of the squares of $\|u\|$ and $\|v\|$; that is:

$$\|u + v\|^2 = \|u\|^2 + \|v\|^2$$

[>

Exercise 1.3: (By hand) Draw the vectors $u = [3, 1]$ and $v = [-2, 6]$. Check that these vectors are orthogonal. Add the vector $u + v$ to your sketch. Recall that for any vector x, $\|x\|^2 = x.x$. Use this fact to check that $\|u + v\|^2 = \|u\|^2 + \|v\|^2$ for the given vectors.

Student Workspace

Answer 1.3

Exercise 1.4: Prove Theorem 1. Hint: Use the rules of algebra above to expand the right side of the identity

$$||u+v||^2 = (u+v).(u+v)$$

Student Workspace

Answer 1.4

[>

Section 2: Projection of a Vector onto a Line

In this section we solve a geometry problem that will have significant applications. The problem is:

- Given a line through the origin and a point not on that line, find the shortest distance from the point to the line. Moreover, find the point on the line that yields this shortest distance.

[>

====================

Example 2A: Given the point (1, 3) and the line $y = x/2$ (see figure below), find the point P on the line that is closest to (1, 3). That is, find P such that the distance from (1, 3) to P is shorter than the distance between (1, 3) and any other point on the line.

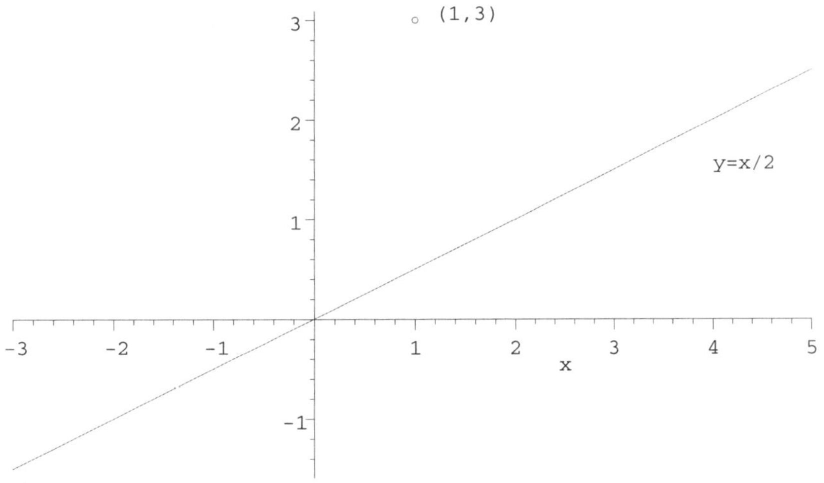

[>

Solution: The point P will lie at the foot of the perpendicular drawn from (1, 3) to the given line. We will use vectors to find P. In the figure below, three vectors have been added to the picture. The vector $u = [1, 3]$ has its head at our given point; the vector $v = [4, 2]$ is a direction vector for the line $y = x/2$; and the blue vector with its head at P is the vector that we wish to calculate exactly. We will call this third vector p.

On the printed copy, draw the vector q from the head of p (i.e., P) to the head of u (i.e., (1, 3)). Note that $q = u - p$ and also that q is orthogonal to v. Thus, $||u - p||$ gives us the shortest distance from (1, 3) to the line.

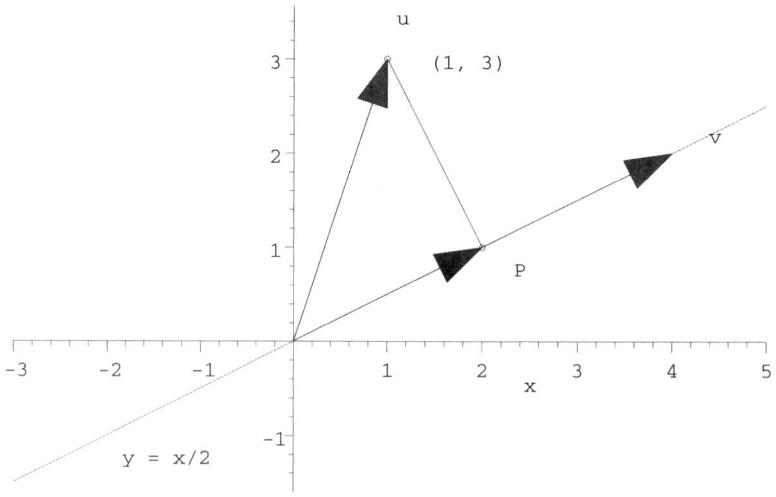


```
[ >
```

The blue vector p is called the *projection* of the vector u onto the vector v. This projection vector p can be calculated using the standard formula:

$$p = \frac{u.v}{v.v}\, v$$

Here is the calculation using Maple:

```
> u := colvector([1,3]);
  v := colvector([4,2]);
> p := evalm(dotproduct(u,v)/dotproduct(v,v)*v);
```

So $P = (2, 1)$, and the distance from (1, 3) to the line is $||u - p|| = \sqrt{5}$:

```
> mag(u-p);
>
```

Note: Instead of choosing v to be [4, 2], we could have chosen v to be any nonzero vector along the line $y = x/2$.

====================

As a consequence of our method for finding the point on a line closest to a given point, we also obtain an "orthogonal decomposition" of the vector u with respect to the line $y = x/2$. That is, we can write u as a sum $u = p + q$, where p is on the line and q is orthogonal to the line:

```
> evalm(p) = c*evalm(v);
```

```
> q := evalm(u-p);
  dotproduct(q,v);
>
```

Exercise 2.1: For the vectors u and v below, find:
(a) The projection p of u onto v.
(b) The shortest distance from the head of u to the line containing v.
(c) Find the orthogonal decomposition of u with respect to the line containing the vector v.

```
> u := colvector([1,1,2]);
  v := colvector([-1,2,1]);
>
```

Student Workspace

Answer 2.1

Exercise 2.2: Derive the formula $p = \frac{u.v}{v.v} v$ for the projection of u onto v by completing this argument:
i) $p = k\,v$ since p is a multiple of v. (So our task is simply to find the scalar k.)
ii) If p is the projection vector, then $u - p$ must be orthogonal to v. Therefore $(u - p).v = 0$, and hence $(u - k\,v).v = 0$.
iii) Expand the final equation from ii) to solve for k.

Student Workspace

Answer 2.2

```
>
```

Maple Note: For convenience we can use the command project(u,v) to find the projection of any vector onto a second vector. For example, using the vectors u and v in Exercise 2.1, we have:

```
> evalm(dotproduct(u,v)/dotproduct(v,v)*v);
> project(u,v);
>
```

Section 3: Projection of a Vector onto a Subspace

In Section 2 we saw how to find the shortest distance from a point to a line by projecting the point onto the line. In this section we introduce the more general idea of projection onto any subspace of R^n, which we will use to solve the following more general shortest distance problem:

- Given a nonzero subspace W of R^n and a point Q not in W, find the shortest distance from Q to W. Moreover, find the point in W that yields this shortest distance.

```
>
```

Before considering this general problem, we consider a more concrete version:

- Given a plane through the origin and a point not on that plane, find the shortest distance from the point to the plane. Moreover, find the point on the plane that yields this shortest distance.

====================

Example 3A: Let u be the vector given below, and let W be the plane spanned by the vectors v_1 and v_2 below. Find the point P on W closest to the point Q at the head of u. That is, find P so the distance from Q to P is shorter than the distance between Q and any other point on the plane. Note: This problem is much easier to solve if the spanning vectors for the plane are orthogonal; so we chose v_1 and v_2 that way.

```
> u := colvector([-4,5,6]);
  v1 := colvector([-3,1,-1]);
  v2 := colvector([1,4,1]);
> 
```

Solution: The point P will lie at the foot of the perpendicular drawn from u to the given plane. The figure below shows the two vectors v_1 and v_2 that span W, the vector u, and the vector p whose head is at the point P. On your printed copy of the figure, label the vectors v_2, u, and p and the points P and Q. Also, add an arrowhead to the top of the black line segment, and label this vector from P to Q as q. So $q = u - p$.

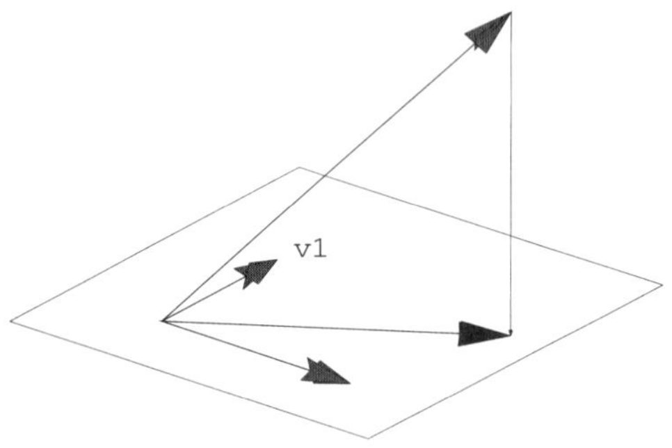


```
> 
```

Note that the vector $q = u - p$ is perpendicular to every vector in W. In fact, of all the vectors in W, only p makes $u - p$ perpendicular to every vector in W. To find this projection vector p, we cannot simply use the projection formula in Section 1, since that gives us the projection onto a line, not a plane. Instead, the formula below, which is proved in Problem 5, does the job. Note that the formula depends on the vectors v_1 and v_2 being orthogonal.

- If W is a plane through the origin in R^3 and v_1, v_2 are orthogonal basis vectors for W, then the projection of any vector u onto W is the sum of the projections of u onto v_1 and u onto v_2:

$$p = \text{project}(u, v_1) + \text{project}(u, v_2)$$

First check that v_1 and v_2 are indeed orthogonal:

```
[ > dotproduct(v1,v2);
```

Calculate the projection vector using the formula from above:

```
[ > p := evalm(project(u,v1)+project(u,v2));
[ >
```

Thus $\left[-\frac{16}{9}, \frac{53}{9}, \frac{2}{9}\right]$ is the exact point in W closest to the head of u.

====================

Exercise 3.1:

(a) Find the shortest distance from the head of u to the plane W spanned by v_1 and v_2 above.

(b) Find the orthogonal decomposition of u with respect to W. That is, express the vector u as the sum of two vectors, where one is a vector in W and the other is a vector orthogonal to W. Check that the second vector is orthogonal to v_1 and v_2 (and hence to every vector in W).

```
[ >
```

Student Workspace

Answer 3.1

Now we return to the general problem posed at the beginning of the section. First we need to define what we mean by a projection of a vector onto a subspace. The idea of the definition is explicit in Example 3A, where we required that the projection p of the vector u must lie in the plane and must have the property that $u - p$ is orthogonal to every vector in the plane. Thus:

> ***Definition:*** If u is any vector in R^n and W is any subspace of R^n, the ***projection of*** u ***onto*** W is the vector p that has the following two properties:
>
> (i) p is in W; and (ii) $u - p$ is orthogonal to every vector in W.

```
[ >
```

In the next module, we will see why the projection p defined above always exists and is uniquely determined by u and W; in the process, we will derive a general formula for p.

The following theorem tells us that the projection vector p solves the general problem posed at the beginning of this section.

> ***Theorem 2:*** If u is any vector in R^n and W is any nonzero subspace of R^n, then the projection p of u onto W is the vector in W that is closest to u. That is, $\|u - p\| < \|u - w\|$ for every vector w in W ($w \neq p$).

[>

Exercise 3.2: (By hand) Prove Theorem 2 by completing this argument:

i) Write the vector $u - w$ as the sum of two vectors: $u - w = (u - p) + (p - w)$.

ii) Show that the two vectors in the above sum, $u - p$ and $p - w$, are orthogonal to each other.

iii) Apply Theorem 1 (Pythagorean Theorem) to the three vectors $u - w$, $u - p$, $p - w$, and deduce the conclusion of Theorem 3 from that formula.

Student Workspace

Answer 3.2

[>

Problems

Problem 1: Distance and angle in the hypercube

The hypercube is the set of points $[x_1, x_2, x_3, x_4]$ in R^4 such that x_1, x_2, x_3, x_4 all range between 0 and 1. So we can think of the hypercube as analogous to the solid unit cube in R^3.

(a) Find the length of longest diagonal of the hypercube, say from the vertex [0, 0, 0, 0] to the vertex [1, 1, 1, 1].

(b) Find the angle between this longest diagonal and any of the edges of the cube, say the edge from [0,0,0,0] to [1,0,0,0]. Recall that the angle θ between two vectors u and v is given by

$$\|u\|\|v\|\cos(\theta) = u.v$$

Student Workspace

Problem 2: Estimate distance from a drawing

On graph paper, draw the vectors $u = [0, -2]$ and $v = [2, 1]$.

(a) Without making any calculation, draw the projection of u onto v and estimate the distance from the head of u to the line containing v.

(b) Check your answer in (a) by calculating the projection and the distance.

Student Workspace

Problem 3: Project a vector on a line

(a) For the vectors x and y below, use the projection of x onto y to find

i) the distance from the head of x to the line containing y;

ii) the orthogonal decomposition $x = p + q$, where p is on the line containing y and q is orthogonal to the line.

```
> x := colvector([2,3,-1]);
  y := colvector([0,-4,2]);
```

[>

(b) Replace x by z, where z is defined below. (z is orthogonal to y.) Answer (i) and (ii) without doing any computation. Explain.

(c) Replace x by w, where w is defined below. (w is parallel to y.) Answer (i) and (ii) without doing any computation. Explain.

```
[ > z := colvector([-2,1,2]);
[ > w := colvector([0,6,-3]);
[ >
```

Student Workspace

Problem 4: Project a vector on a plane

(a) Consider the plane W given by $2x + y + 3z = 0$ and the vector $u = [5, 2, 10]$. Use the projection of u onto W to find

i) the distance from the head of u to the plane W;

ii) the decomposition $u = p + q$, where p is in W and q is orthogonal to every vector in W.

Hint: You will have to find a pair of orthogonal vectors v_1, v_2 that span W. First find any two spanning vectors, and then use the method in Section 2 to construct an orthogonal pair from your spanning vectors.

(b) Note that the vector q is parallel to the vector $[2, 1, 3]$ formed by the coefficients of x, y, z in the equation of the plane. Why is this true?

Student Workspace

Problem 5: Orthogonal basis vectors and projections

We solved the shortest distance problem in Section 3 by using the following result:

If W is a plane through the origin in R^3 and v_1, v_2 are orthogonal basis vectors for W, then the projection p of any vector u onto W is the sum of the projections of u onto v_1 and u onto v_2:

$$p = \text{project}(u, v_1) + \text{project}(u, v_2)$$

Derive this formula for the projection of u onto W by completing this argument:

i) Since the projection vector p is in W, we can express p as a linear combination of the basis vectors $\{v_1, v_2\}$: $p = a_1 v_1 + a_2 v_2$. (So our task is to find the scalar weights a_1 and a_2.)

ii) In order for p to be the projection vector of u onto W, $u - p$ must be orthogonal to v_1 and v_2. Therefore $(u - a_1 v_1 - a_2 v_2).v_1=0$ and $(u - a_1 v_1 - a_2 v_2).v_2=0$

iii) Expand the two expressions in ii) to find a_1 and a_2.

Student Workspace

Problem 6: Construct an orthogonal basis of R^3

Use projections to find a basis $\{v_1, v_2, v_3\}$ of R^3, where $v_1 = [2, 5, -4]$ and each of the vectors v_1, v_2, v_3 is orthogonal to the other two. (Such a basis is called an "orthogonal basis" of R^3.) Hint: Notice that the set of vectors $\{v_1, v_2, q\}$ in Problem 4 is an orthogonal basis of R^3; so similar techniques should work.

Student Workspace

Problem 7: Projection transformations

If v is any nonzero vector in R^n, then the matrix P that projects every vector in R^n onto the line through the origin with direction vector v is given by the formula $P = \frac{v\,v^T}{v^T\,v}$. Here's the explanation:

For any vector u in R^n, its projection $P\,u$ onto the line through the origin with direction vector v is $\frac{u.v}{v.v}\,v$. Rewrite this formula using the fact that $u.v = v.u = v^T\,u$ and $v.v = v^T\,v$. Also, put the scalar $\frac{u.v}{v.v}$ on the right of the vector v. Thus, $P\,u = \frac{v\,v^T\,u}{v^T\,v}$, and therefore the matrix P is $\frac{v\,v^T}{v^T\,v}$.

(a) Use the above formula to find the 2 by 2 matrix that projects every vector in R^2 onto the line $y = x$. (Do not use angles. However, you may check your answer against the matrix projectmat(π/4).)
(b) Use the above formula to find the 3 by 3 matrix that projects every vector in R^3 onto the line through the origin with direction vector $[2, 1, -3]$.
(c) By a derivation similar to the one above, find the 3 by 3 matrix that projects every vector in R^3 onto the plane with normal vector $[2, 1, -3]$. (Hint: If P is the desired projection matrix and u is an arbitrary vector in R^3, then $P\,u = u - p$, where p is the projection of u onto the normal vector.) You may check your answer against the matrix projectmat3d([2,1,-3]).

Student Workspace

Paper and Pencil Practice Problems

Note: Maple is not required for any of the problems in this section. Do all of these by hand.

1. Let $u = [1, 4]$ and $v = [3, 1]$.
(a) Find the projection of u onto v.
(b) Find the distance from the head of u to the line containing v.
(c) Draw a picture that illustrates your method of solution.

2. Let W = Span$\{v_1, v_2\}$, where $v_1 = [1, 0, -1]$ and $v_2 = [2, 1, 0]$.
(a) Find a pair of orthogonal vectors that span W. (Try to answer this without a calculation.)
(b) Find the projection of $u = [2, 0, 1]$ onto W.

Linear Algebra Modules Project
Chapter 7, Module 2

Orthogonal Bases

Purpose of this module

The purpose of this module is to see how to construct and make use of an orthogonal basis of a subspace. In particular, we see how to find the projection of any vector on any subspace and thus find the shortest distance from the vector to the subspace. We also use orthogonal bases in developing the concepts of orthogonal complement and orthogonal matrix.

Prerequisites

Dot product of vectors, orthogonal vectors, length of a vector, projection of one vector on another; basis of a subspace; null space and row space of a matrix; matrix transformations.

Commands used in this module

```
> restart; with(linalg): with(lamp):
>
```

Tutorial

Section 1: Definition and Properties of Orthogonal Bases

We begin with two definitions:

Definition: A set of vectors $\{v_1, ..., v_k\}$ in R^n is called an ***orthogonal set*** if every pair of vectors in the set is orthogonal.

Definition: An ***orthogonal basis*** of a subspace S is a basis of S which is also an orthogonal set. That is, every pair of vectors in the basis is orthogonal.

The standard basis of R^n is an orthogonal basis of R^n. We confirm this for the standard basis of R^3:

```
> e1 := colvector([1,0,0]);
  e2 := colvector([0,1,0]);
  e3 := colvector([0,0,1]);
> dotproduct(e1,e2);
> dotproduct(e1,e3);
> dotproduct(e2,e3);
>
```

Exercise 1.1: Find an orthogonal basis of the subspace of R^4 in which one of the vectors is [1, 1, 1, 0]. (This can be done without any computation.)

Student Workspace

Answer 1.1

The following theorem shows that any linear combination of orthogonal vectors can be thought of as a sum of projections.

[>

Theorem 3: If $\{ v_1, ..., v_k \}$ is an orthogonal basis of a subspace S of R^n and u is any vector in S, then u is the sum of its projections on $v_1, ..., v_k$. Thus, the coordinates of u relative to this basis are $u.v_1/v_1.v_1, \; ..., \; u.v_k/v_k.v_k$.

Proof: Since u is in the span of $\{ v_1, ..., v_k \}$, it is a linear combination of $v_1, ..., v_k$:

$$u = c_1\, v_1 + ... + c_k\, v_k \qquad [1]$$

Take the dot product of each side of equation [1] with v_1:

$$u.v_1 = c_1\, v_1.v_1 + c_2\, v_2.v_1 + . . . + c_k\, v_k.v_1$$

Since $\{ v_1, ..., v_k \}$ is an orthogonal set, every term on the right side is zero, except for the first term. Solving $u.v_1 = c_1\, v_1.v_1$ for c_1, we get $c_1 = u.v_1/v_1.v_1$, which shows that the first term on the right side of equation [1], $c_1\, v_1$, is the projection of u onto v_1. Similarly, by taking the dot product of each side of equation [1] with $v_2, ..., v_k$, we see that each of the remaining terms on the right side of equation [1] also is a projection.

Here's a useful theorem that we can see is plausible by visualizing, say, three vectors in R^3:

[>

Theorem 4: If $\{ v_1, ..., v_k \}$ is an orthogonal set of nonzero vectors in R^n, then $\{ v_1, ..., v_k \}$ is a linearly independent set of vectors.

Exercise 1.2: Prove Theorem 4. Hint: You must show that the only solution to

$$0 = c_1\, v_1 + ... + c_k\, v_k$$

is the trivial solution in which all the coefficients are zero. Note that this equation has the form of equation [1], and use Theorem 3.

Student Workspace

Answer 1.2

[>

Section 2: Orthogonal Projections onto Orthogonal Bases

In this section we will use projections onto an orthogonal basis to solve the following problem:

- Given a nonzero subspace W of R^n and a point Q not in W, find the shortest distance from Q to W. Moreover, find the point in W that yields this shortest distance.

According to Theorem 2 (see Module 1), we can solve this problem by finding the projection P of Q onto W. The shortest distance from Q to W is simply the distance between Q and P. Furthermore, in Example 3A of Module 1, we saw that if W has dimension 2, we can find the projection of any vector u onto W by adding the projections of u onto an orthogonal basis of W. Let's review Example 3A briefly:

=====================

Example 2A: Let u be the vector given below, and let W be the plane spanned by the vectors v_1 and v_2 below. Find the point on W closest to the head of u.

```
> u := colvector([-4,5,6]);
  v1 := colvector([-3,1,-1]);
  v2 := colvector([1,4,1]);
```

[>

Solution: The figure below shows the two vectors v_1 and v_2 that span W, the vector u, and the projection p of the vector u onto W.

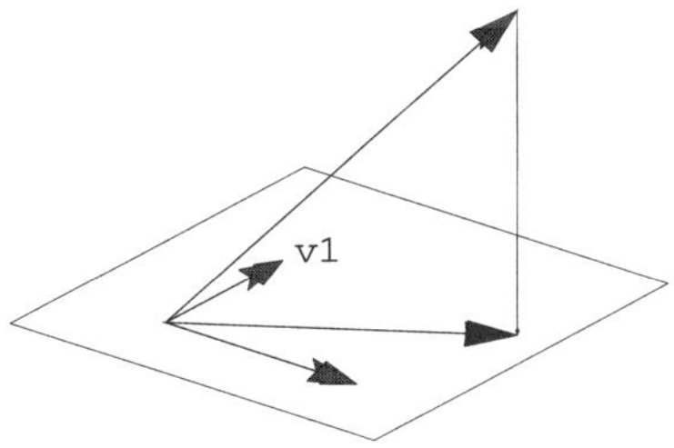

[>

Since p is the projection of u onto W, the vector $q = u - p$ is perpendicular to every vector in W and therefore to v_1 and v_2. We have chosen v_1, v_2 to be orthogonal, which turns out to be crucial in finding the projection of u onto the plane W; we simply add the projections of u onto v_1 and u onto v_2:

$$p = \text{project}(u, v_1) + \text{project}(u, v_2)$$

```
[ > p := evalm(project(u,v1)+project(u,v2));
[ >
```

Thus $\left[-\frac{16}{9}, \frac{53}{9}, \frac{2}{9}\right]$ is the point in W closest to the head of u.

===================

The next theorem generalizes the above result, that the projection p of u is the sum of the projections of u onto v_1 and v_2:

Theorem 5: If { v_1, ..., v_k } is an orthogonal basis of a subspace W of R^n and u is any vector in R^n, then the projection p of u onto W is the sum of the projections of p onto v_1, ..., v_k:

$$p = \text{project}(u, v_1) + \ldots + \text{project}(u, v_k)$$

```
[ >
```

To prove this result, we must show that the above formula for p as a sum of projections has the two properties of a projection:

(i) p is in W; and (ii) $u - p$ is orthogonal to every vector in W.

Since each of the projections is in W, their sum is in W. So we need only verify property (ii), which is left to Problem 6.

Here's an example of the above method:

====================

Example 2B: Find the projection p of u onto the subspace W with the orthogonal basis { v_1, v_2, v_3 } (see below). Also find the shortest distance from the head of u to W.

```
[ > u := colvector([1,0,-2,1,3]);
[ > v1 := colvector([0,1,-1,0,0]);
    v2 := colvector([2,0,0,1,0]);
    v3 := colvector([1,1,1,-2,3]);
```

Solution: We compute the sum of the projections of u onto v_1, v_2, v_3 to find p; then $\|u - p\|$ is the shortest distance:

```
[ > p := evalm(project(u,v1)+project(u,v2)+project(u,v3));
[ > q := evalm(u-p):
    mag(q);
```

So the distance from the head of u to W is $\sqrt{895/10}$. We can also check that q is orthogonal to W:

```
[ > dotproduct(q,v1); dotproduct(q,v2); dotproduct(q,v3);
```

```
[ >
```

====================

Section 3: Gram-Schmidt Process

The *Gram-Schmidt process* is a method for constructing an orthogonal basis of a subspace from any given basis. That is, suppose we are given a basis $\{u_1, ..., u_k\}$ of a nonzero subspace S of R^n. We will use this basis to construct an orthogonal basis of S. Let's look at an example first.

====================

Example 3A: The vectors u_1, u_2 below span a two-dimensional subspace S of R^3.

```
> u1 := colvector([1,1,2]);
> u2 := colvector([-1,2,1]);
[ >
```

We will construct an orthogonal basis of S by again using a projection. We choose u_1 itself to be the first vector in our orthogonal basis. Then we find the projection p of u_2 onto u_1. Since the vector $q = u_2 - p$ is orthogonal to u_1, $\{ u_1, q \}$ is an orthogonal basis of S.

```
[ > p := project(u2,u1);
[ > q := evalm(u2-p);
```

Let's check that u_1 and q are indeed orthogonal:

```
[ > dotproduct(u1,q);
```

Here's a picture that shows both the original basis $\{ u_1, u_2 \}$ and the orthogonal basis $\{ u_1, q \}$:

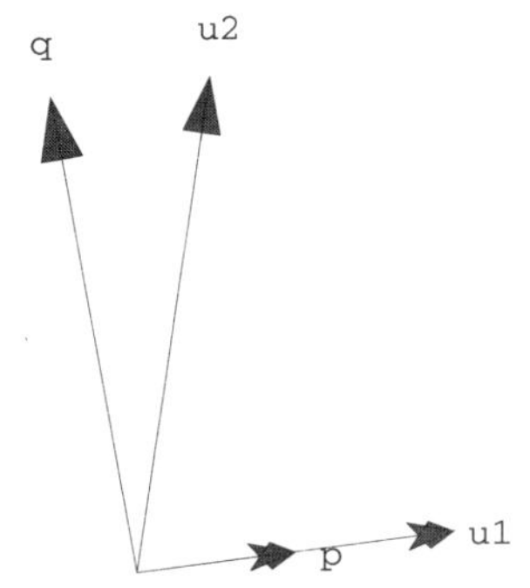


```
[ >
```

====================

Now suppose we have a 3-dimensional subspace S of R^n and a basis $\{ u_1, u_2, u_3 \}$ of S. To construct an orthogonal basis $\{ v_1, v_2, v_3 \}$ of S, we start just as we did in Example 3A: We choose v_1 to be u_1

and choose v_2 to be $u_2 - p_1$, where p_1 is the projection of u_2 onto v_1. To find v_3, we project u_3 onto the subspace spanned by the orthogonal vectors v_1 and v_2. That is, we construct

$$p_2 = \text{project}(u_3, v_1) + \text{project}(u_3, v_2)$$

We then choose v_3 to be $u_3 - p_2$.

Try this method in the following exercise:

Exercise 3.1: The vectors u_1, u_2, u_3 below span a three-dimensional subspace S of R^4. Use the above method to find an orthogonal basis { v_1, v_2, v_3 } for S.

```
> u1 := colvector([1,1,2,0]);
  u2 := colvector([-1,2,1,0]);
  u3 := colvector([1,0,0,1]);
>
```

Student Workspace

Answer 3.1

Here is a step-by-step description of the Gram-Schmidt process for converting any basis { u_1, ..., u_k } of a subspace S of R^n to an orthogonal basis { v_1, ..., v_k } of S:

1. Choose v_1 to be u_1.
2. Choose v_2 to be u_2 - project(u_2, v_1).
3. Choose v_3 to be u_3 - (project(u_3, v_1) + project(u_3, v_2)).

and so on. That is, at each step, the new vector v_j is chosen to be u_j minus the projection of u_j on the subspace spanned by the preceding basis vectors. This projection is found by adding the projections of u_j onto each of the preceding basis vectors, provided we project u_j onto orthogonal basis vectors.

In particular, we have proved the following theorem:

Theorem 6: Every nonzero subspace of R^n has an orthogonal basis.

```
>
```

Section 4: Orthogonal Complements

Definition: Suppose S is a subspace of R^n. The ***orthogonal complement*** of S in R^n is the set of all vectors v in R^n such that $v.u = 0$ for all u in S, that is, the set of all vectors that are orthogonal to every vector in S.

=====================

Example 4A: The orthogonal complement of the line $y = 2\,x$ in R^2 is $y = -x/2$, the perpendicular line through the origin.

=====================

Exercise 4.1: What is the orthogonal complement of the plane $2x + y - 3z = 0$ in R^3? (No computation is needed.)

Student Workspace

Answer 4.1

```
>
```

Exercise 4.2: What is the orthogonal complement of the orthogonal complement you found in Exercise 4.1?

Student Workspace

Answer 4.2

We can now use the fact that subspaces have orthogonal bases to prove the following theorem:

Theorem 7: If S_1 is a subspace of R^n and S_2 is its orthogonal complement in R^n, then

$$\text{dimension}(S_1) + \text{dimension}(S_2) = n$$

```
>
```

Sketch of Proof of Theorem 7

By Theorem 6, we can construct orthogonal bases $\{ u_1, ..., u_k \}$ and $\{ v_1, ..., v_m \}$ for S_1 and S_2, respectively. Since k is the dimension of S_1 and m is the dimension of S_2, we must prove that $k + m = n$. By Theorem 4, the combined set of $k + m$ vectors, $\{ u_1, ..., u_k, v_1, ..., v_m \}$, is linearly independent. Furthermore, if w is any vector in R^n, it is not hard to show that w equals its projection onto the span of $\{ u_1, ..., u_k, v_1, ..., v_m \}$. Therefore $\{ u_1, ..., u_k, v_1, ..., v_m \}$ spans R^n and hence is a basis of R^n. Since any basis of R^n has exactly n vectors, we have proved that $k + m = n$.

```
>
```

Row Space and Column Space Revisited

The concept of orthogonal complement gives us additional geometric insight into the relationship between the null space and row space of a matrix. The following example will give the idea.

====================

Example 4B: For the matrix A below, find a basis for the null space and compute the dot product of each null space basis vector with each row vector of A:

```
> A := matrix([[0,1,-1],[-1,-4,2]]);
```

Solution:

```
> N := nullbasis(A);
> N1 := N[1];
```

```
> r1 := row(A,1);
> r2 := row(A,2);
> dotproduct(r1,N1);
> dotproduct(r2,N1);
>
```

====================

This orthogonality of the vectors is not a coincidence. A vector x is in the null space of a matrix A if and only if $A\,x = 0$. Since $A\,x$ is computed by taking the dot product of each row of A with x, we see that x is in the null space of A if and only if x is orthogonal to each row of A. Therefore:

Theorem 8: The orthogonal complement of the row space of A is the null space of A.

Note that we now have a new proof of Theorem 9 of Chapter 5, which says that the dimension of the null space of A is $n - r$, where n is the number of columns of A and r is the rank of A. Proof: The dimension of the row space of A is r, and so by Theorem 7 the dimension of its orthogonal complement is $n - r$. By Theorem 8 this orthogonal complement is the null space of A.

```
>
```

Section 5: Orthogonal Matrices

Every rotation matrix preserves the angle between any two vectors and preserves the length of any vector. For example, the vectors u and v below are orthogonal, and the rotation by the angle $\pi/3$ sends u and v to output vectors that are also orthogonal and whose lengths are the same as u and v, respectively. We confirm these facts both algebraically and geometrically:

```
> u := colvector([1,0]);
  v := colvector([0,2]);
> R := rotatemat(Pi/3);
> dotproduct(R&*u,R&*v);
  mag(R&*u);
  mag(R&*v);
> drawvec2d(u,v,[R&*u,headcolor=blue],[R&*v,headcolor=blue]);
>
```

Reflection matrices also preserve angle and length. Another interesting (and, as we will see, related) property of rotation and reflection matrices is that the product of the matrix and its transpose is the identity matrix:

```
> R := rotatemat(theta);
> RTR := evalm(transpose(R)&*R);
> simplify(RTR);
> S := reflectmat(theta);
> STS := evalm(transpose(S)&*S);
> simplify(STS);
>
```

These properties hold for a large class of interesting matrix transformations, which are defined as follows:

Definition: A square matrix U is ***orthogonal*** if $U^T\,U = I$.

Another way to state this definition is that $U^{(-1)} = U^T$ or, equivalently, $U\,U^T = I$. The following exercise will indicate yet another way to restate the definition of orthogonality.

Exercise 5.1: (a) (By hand) Show that the columns of the matrix R = rotatemat($\pi/3$) (see below) are orthogonal and have length 1.
(b) (By hand) With the help of part (a), show that $R^T\,R = I$.

$$R = \begin{bmatrix} \frac{1}{2} & -\frac{1\sqrt{3}}{2} \\ \frac{1\sqrt{3}}{2} & \frac{1}{2} \end{bmatrix}$$

[>

Student Workspace

Answer 5.1

[>

As you can see from Exercise 5.1, the statement $U^T\,U = I$ is equivalent to the statement that the columns of U form an orthogonal set of unit vectors. Such a set of vectors is called an *orthonormal set.*

Yet another consequence of the definition of orthogonality is that an orthogonal matrix preserves dot products. Here's the proof:

$$U\,x.U\,y = (U\,x)^T\,U\,y = x^T\,U^T\,U\,y = x^T\,y = x.y$$

From this conclusion we can show that all orthogonal matrices preserve angles and lengths. (See Exercise 5.2 and Problem 9.)

Exercise 5.2: Use the above formula, $U\,x.U\,y = x.y$, to show that orthogonal matrices preserve lengths of vectors.

Student Workspace

Answer 5.2

In fact, the various properties we have been discussing are all equivalent to one another, as the following theorem says.

[>

Theorem 9: Let U be an n by n matrix. The statement that U is orthogonal (i.e., $U^T\,U = I$) is equivalent to each of the following statements:

(a) $U^{(-1)} = U^T$.

(b) $U\,U^T = I$.

(c) The columns of U form an orthonormal basis of R^n.

(d) The rows of U form an orthonormal basis of R^n.

(e) $U\,x.U\,y = x.y$ for all vectors x and y in R^n.

```
>
```

Problems

Problem 1: Gram-Schmidt applied to three vectors

(a) Using the Gram-Schmidt process (but not the gram_schmidt(..) command), find an orthogonal basis of the subspace of R^5 spanned by the following three linearly independent vectors:

```
> u1 := colvector([1,0,2,0,-1]);
  u2 := colvector([0,1,1,-2,0]);
  u3 := colvector([-1,2,0,1,-3]);
>
```

(b) Express each of the original basis vectors u_1, u_2, u_3 as a linear combination of your orthogonal basis vectors. Hint: This is most easily done by looking at the equations used in the Gram-Schmidt process. No computation should be necessary.

Student Workspace

Problem 2: Gram-Schmidt applied to same vectors, different order

(a) Using the Gram-Schmidt process (but not the gram_schmidt(..) command), find an orthogonal basis of the subspace of R^5 spanned by the following three linearly independent vectors:

```
> u1 := colvector([-1,2,0,1,-3]);
  u2 := colvector([1,0,2,0,-1]);
  u3 := colvector([0,1,1,-2,0]);
>
```

(b) These vectors are the same vectors as in Problem 1. Why did we get a different orthogonal basis than in Problem 1?

Student Workspace

Problem 3: Gram-Schmidt applied to linearly dependent vectors

(a) Using the Gram-Schmidt process (but not the gram_schmidt(..) command), find an orthogonal basis of the subspace of R^5 spanned by the following four vectors:

```
> u1 := colvector([1,0,2,0,-1]);
  u2 := colvector([0,1,1,-2,0]);
```

```
   u3 := colvector([1,-1,1,2,-1]);
   u4 := colvector([-1,2,0,1,-3]);
> 
```

(b) Check your answer using the gram_schmidt(..) command.

(c) Something quite different happened in this problem than in Problems 1 or 2. What happened and why did it happen? Explain geometrically. (Hint: Notice that u_3 is a linear combination of u_1 and u_2. Compare your orthogonal basis in this problem with your orthogonal basis in Problem 1.)

Student Workspace

Problem 4: Distance from a point to a subspace

Find the projection of the vector u onto the subspace S spanned by v_1, v_2, v_3, v_4 (see below). Also find the shortest distance from the head of u to the subspace S. (You may use the gram_schmidt(..) command.)

```
> u := colvector([1,0,0,0,0]);
> v1 := colvector([1, 3, 1, 0, 3]);
  v2 := colvector([3, -5, 1, 2, 5]);
  v3 := colvector([-5, 1, -2, 3, -1]);
  v4 := colvector([0, -4, 4, 1, -3]);
> 
```

Student Workspace

Problem 5: Orthogonal bases for null space and row space

(a) Find an orthogonal basis for the null space of the matrix A below. (You may use the gram_schmidt(..) command.)

```
> A := matrix([[2,-3,-4,-4,-3],[-9,4,-1,-1,4],[-5,1,-3,-3, 1]]);
> 
```

(b) Find an orthogonal basis for the orthogonal complement of the null space of A. (You may use the gram_schmidt(..) command.)

(c) Check that the combined set of the bases in (a) and (b) is linearly independent. Why is this combined set an orthogonal basis of R^5?

Student Workspace

Problem 6: Defining property of a projection

If $\{ v_1, v_2, v_3 \}$ is an orthogonal set of vectors in R^n and u is any vector in R^n and

$$p = \text{project}(u, v_1) + \text{project}(u, v_2) + \text{project}(u, v_3)$$

prove that $u - p$ is orthogonal to each of the vectors v_1, v_2, v_3.

Student Workspace

Problem 7: Product of orthogonal matrices

If U and V are n by n orthogonal matrices, prove, using only the definition of orthogonality, that the product $U\,V$ is orthogonal.

Student Workspace

Problem 8: Reflections in R^3

(a) Verify that every 3 by 3 reflection matrix is orthogonal. Use the command below to construct these matrices. (The vector $[a, b, c]$ is a normal vector to the plane of reflection.)

```
> R := reflectmat3d([a,b,c]);
>
```

(b) According to Problem 7, the product of the two reflection matrices S and T below is also orthogonal. What familiar matrix does the product $S\,T$ equal? Use geometric reasoning to determine your answer. Also, check your answer by computing the product and comparing it with your answer.

```
> S := reflectmat3d([0,0,1]);
  T := reflectmat3d([0,1,0]);
>
```

Student Workspace

Problem 9: Orthogonal matrices preserve angles

If U is any n by n orthogonal matrix and x and y are any two vectors in R^n, prove that the angle between x and y equals the angle betwen $U\,x$ and $U\,y$. Hint: Use the formula relating dot product and angles:

$$x.y = ||x||\,||y||\cos(\theta)$$

Student Workspace

Pencil and Paper Practice Problems

Note: Maple is not required for any of the problems in this section. Do all of these by hand.

1. Find an orthogonal basis for Span$\{u_1, u_2, u_3\}$ (see below). Note that u_1 and u_2 are already orthogonal.

$$u_1 = \begin{bmatrix} 2 \\ -3 \\ 0 \\ 1 \end{bmatrix}, \quad u_2 = \begin{bmatrix} 3 \\ 1 \\ 4 \\ -3 \end{bmatrix}, \quad u_3 = \begin{bmatrix} 0 \\ 2 \\ 4 \\ -1 \end{bmatrix}$$

2. (a) Find the orthogonal complement in R^3 of the line

$$x = 2\,t, \quad y = -3\,t, \quad z = 4\,t$$

(b) Find a basis $\{ u_1 \}$ for the line and an orthogonal basis $\{ u_2, u_3 \}$ for the orthogonal complement.

(c) Explain how we know, without any computation, that $\{ u_1, u_2, u_3 \}$ is an orthogonal basis of R^3.

Linear Algebra Modules Project
Chapter 7, Module 3

Least Squares Solutions

Purpose of this module

The purpose of this module is to introduce the idea of a "least-squares solution" of a linear system that has no solutions. This has applications to problems in curve fitting. One such example is the problem of fitting a straight line to a set of data points that are nearly linear.

Prerequisites

Linear systems; null space and column space of a matrix; orthogonal vectors, projections, orthogonal complements.

Commands used in this module

```
> restart; with(linalg): with(lamp): with(plots):
>
```

Tutorial

Section 1. Introduction

In real-world problems, we sometimes encounter a system of linear equations $A\,x = b$ that has no solutions but for which an "approximate" solution would be useful. (We will see examples of such problems in Section 3.) For instance, the following linear system has no solutions:

$$x_1 + 2\,x_2 = 1$$
$$3\,x_1 + 5\,x_2 = 3$$
$$-x_1 + x_2 = 2$$

====================

Example 1A: We confirm that the system has no solutions by using matsolve(..):

```
> A := matrix([[1,2],[3,5],[-1,1]]);
  b := colvector([1,3,2]);
> x := matsolve(A,b);
>
```

====================

Although $A\,x = b$ has no solution, can we find the "best approximate solution" to $A\,x = b$? The criterion we will use in determining such a "solution" is to require that $A\,x$ be as close as possible to b. Another way to phrase this is to require that the distance from $A\,x$ to b, which is $\|A\,x - b\|$, be as small as possible. Thus, our criterion is:

- The "best approximate solution" to $A\,x = b$ is a vector x such that $\|A\,x - b\|$ is as small as possible.

====================

Example 1B: Which of the two vectors x_1 and x_2 defined below is a better approximate solution to $A\,x = b$, according to the above criterion?

```
> x1 := colvector([2,1]);
> x2 := colvector([1,-1]);
```

We compute the error, $A\,x - b$, for each of these vectors:

```
> error1 := evalm(A&*x1-b);
  error2 := evalm(A&*x2-b);
```

Then we find the lengths of these vectors and compute their decimal approximations:

```
> evalf(mag(error1));
  evalf(mag(error2));
> 
```

====================

So, according to the above criterion, x_2 is a better approximate solution than x_1. However, is there a best x -- one for which $\|A\,x - b\|$ is smallest? Is there more than one such vector x? And if there is a best choice for x, how do we find it? These three questions are addressed in the next section.

The proper terminology for a "best approximate solution" is a "least-squares solution." Here's why this phrase is used. When we are finding such a "solution" x, we are finding the smallest (and hence "least") value of $\|A\,x - b\|$. For example, the vector $A\,x - b$ in Example 1A is

$$\begin{bmatrix} 1 & 2 \\ 3 & 5 \\ -1 & 1 \end{bmatrix} \begin{bmatrix} x_1 \\ x_2 \end{bmatrix} - \begin{bmatrix} 1 \\ 3 \\ 2 \end{bmatrix} = \begin{bmatrix} x_1 + 2\,x_2 - 1 \\ 3\,x_1 + 5\,x_2 - 3 \\ -x_1 + x_2 - 2 \end{bmatrix}$$

and therefore the value of $\|A\,x - b\|$ is the square root of

$$(x_1 + 2\,x_2 - 1)^2 + (3\,x_1 + 5\,x_2 - 3)^2 + (-x_1 + x_2 - 2)^2$$

So we are finding the least value of this sum of squares; hence the phrase "least-squares."

Definition: A ***least-squares solution*** to a linear system $A\,x = b$ is a vector x that yields the smallest value of $\|A\,x - b\|$.

[>

Section 2. Least-Squares Solutions and the Normal Equations

Projections are the key to finding a least-squares solution of $A\,x = b$. Recall that the product $A\,x$ can be thought of as a linear combination of the columns of A:

$$\begin{bmatrix} 1 & 2 \\ 3 & 5 \\ -1 & 1 \end{bmatrix} \begin{bmatrix} x_1 \\ x_2 \end{bmatrix} = x_1 \begin{bmatrix} 1 \\ 3 \\ -1 \end{bmatrix} + x_2 \begin{bmatrix} 2 \\ 5 \\ 1 \end{bmatrix}$$

[>

So the vector $A\,x$ is a vector in the column space of A. Therefore $A\,x = b$ has a solution if and only if b is in the column space of A. We are interested in the case when b is <u>not</u> in the column space of A. In this case, we want to find x so that the distance from b to $A\,x$ (and hence the distance to the column space of A) is as small as possible. As we saw in Modules 1 and 2, the shortest distance from the head of a vector b to a subspace W is found by projecting b onto W. Here's the picture:

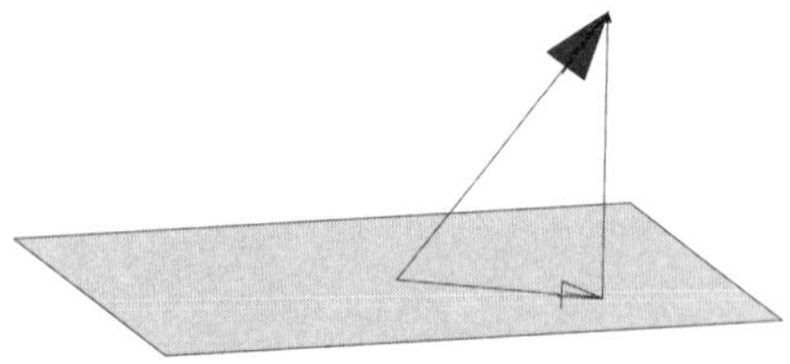

[>

The plane represents the column space of A; the vector not lying in the plane is the vector b; the vector in the plane is the projection of b onto the plane; and the black line segment connecting their heads is perpendicular to the plane.

This discussion suggests the following procedure for finding a least-squares solution of $A\,x = b$:

- 1. Find the projection p of the right-side vector b onto the column space of A. (Recall that we find the projection of a vector b onto a subspace W by finding an orthogonal basis of W, projecting b onto each basis vector, and adding these projections.)
- 2. Solve $A\,x = p$.

Exercise 2.1: Use the above procedure to find the least-squares solution of the linear system $A\,x = b$, where A and b are defined below. Hint: The columns of A are orthogonal, which simplifies the job of finding the projection p.

```
> A := matrix([[1,1],[2,-2],[3,1]]);
  b := colvector([3,4,2]);
```

Student Workspace

Answer 2.1

If the columns of A are not orthogonal, we can use projections to replace them with an orthogonal basis of the column space of A, as in the next example.

Example 2A: Find the least-squares solution of $A\,x = b$, where A and b are as in Example 1A:

```
> A := matrix([[1,2],[3,5],[-1,1]]);
  b := colvector([1,3,2]);
```

Solution: We orthogonalize the columns of A (see Section 3 in Module 2):

```
> u := column(A,1);
  v := column(A,2);
> w := evalm(v-project(v,u));
```

So $\{\,u, w\,\}$ is an orthogonal basis of the column space of A. Then we find the projection of b onto the column space by adding the projections of b onto the orthogonal basis vectors:

```
> p := evalm(project(b,u) + project(b,w));
```

Finally, we solve $A\,x = p$:

```
> x := matsolve(A,p);
>
```

This method is a bit tedious, and it is much more tedious for larger systems. We could speed it up by using the gram_schmidt(..) command to compute the orthogonal basis, but that would still be quite tedious for large systems. Fortunately, there is a much faster method, which we describe next.

The Normal Equations

First let's recall the defining properties of a projection: The projection of a vector b onto a subspace W is the vector p that has the following two properties:

(i) p is in W; and (ii) $b - p$ is orthogonal to every vector in W.

```
>
```

So if W is the column space of a matrix A, property (i) tells us that $p = A\,x$ for some vector x. Also, property (ii) says that the vector $b - p$ is orthogonal to every vector in the column space of A. Since the rows of A^T are the columns of A, we can rephrase this last sentence to say that $b - p$ is orthogonal to every vector in the row space of A^T. In other words, $b - p$ is in the orthogonal complement of the row space of A^T, which is the null space of A^T (by Theorem 8 in Module 2). Thus,

$$A^T\,(b - p) = 0$$

Next, multiply out, and replace p by $A\,x$:

$$A^T\,b - A^T\,A\,x = 0$$

Finally, rearrange terms:

$$A^T\,A\,x = A^T\,b$$

```
>
```

The equations of the linear system $A^T A\,x = A^T b$ are called the *normal equations* for the original linear system $A\,x = b$. Note that the normal equations can be found simply by multiplying the original linear system on the left by A^T. So here is our new, faster method for computing the least-squares solution of $A\,x = b$:

- 1. Multiply the equation $A\,x = b$ on the left by A^T. (The resulting system, $A^T A\,x = A^T b$, is the system of normal equations for the original system.)
- 2. Solve $A^T A\,x = A^T b$. (The normal equations always have a solution.)

Exercise 2.2: (a) Use the normal equations to find the least-squares solution x of the linear system in Example 1A (repeated below). (b) Verify that $A\,x - b$ is orthogonal to every vector in the column space of A.

```
> A := matrix([[1,2],[3,5],[-1,1]]);
  b := colvector([1,3,2]);
>
```

Student Workspace

Answer 2.2

Uniqueness of the Least-Squares Solution

We have seen the answer to two of the three basic questions raised at the end of Section 1: Every linear system $A\,x = b$ has a least-squares solution; and we can find such a solution by solving the normal equations $A^T A\,x = A^T b$. The remaining question is whether there can be more than one least-squares solution to a linear system.

In fact, there *can* be more than one least-squares solution to a linear system. Try this example:

```
> A := matrix([[1,2],[3,6],[-1,-2]]);
  b := colvector([1,2,-1]);
```

We construct and solve the normal equations:

```
> M := evalm(transpose(A)&*A);
  k := evalm(transpose(A)&*b);
> matsolve(M,k);
>
```

Notice that the columns of A are linearly dependent and hence A has a nonzero null space. Whenever a matrix A has a nonzero null space, the linear system $A\,x = b$ has more than one least-squares solution for the following reason. If x is a least-squares solution and n is a nonzero vector in the null space of A, then $A\,(x + n) = A\,x + A\,n = A\,x$, since $A\,n = 0$. So $x + n$ is another least-squares solution of $A\,x = b$. Conversely, if the null space of A is $\{0\}$, the least-squares solution is unique. The proof of this statement is left to Problem 4. In summary:

Theorem 10: Every linear system $A\,x = b$ has a least-squares solution x, and the least-squares solution is unique if and only if the null space of A is $\{0\}$.

```
>
```

Section 3: Fitting Curves to Data

When a scientist has collected some data, one of her first tasks is to find a pattern or a sense of order in the data. To this end, she will usually plot the data, see if the plot seems to follow a familiar shape such as a straight line, parabola, or graph of an exponential or log function, and find the function of that type which best fits the data. The most widely used method for selecting the "best fitting" function of a given type to a set of data points is the least-squares method studied in this section.

====================

Example 3A: Here is a set of data points that nearly lie on a straight line:

```
> data := [[1,3],[2,3],[3,4],[4,5],[5,7]];
> p1 := plot([data],style=POINT,symbol=CIRCLE,color=red):
> display(p1,view=[-1..6,0..7]);
>
```

Solution: We want to find the line $y = m\,t + c$ that best fits the data in a least-squares sense.

```
> f := t -> m*t+c;
```

If the above five points were to lie on this line, we would have the following five equations:

```
> eqn1 := f(1)=3;
> eqn2 := f(2)=3;
> eqn3 := f(3)=4;
> eqn4 := f(4)=5;
> eqn5 := f(5)=7;
```

This system of five linear equations in the two unknowns c and m has no solution, since the five points do not lie on a line. Instead, we will find a least-squares solution. First we write our system of five linear equations as a matrix-vector equation $A\,x = b$:

```
> M := genmatrix([eqn1,eqn2,eqn3,eqn4,eqn5],[c,m],flag);
> A := submatrix(M,1..5,1..2);
  b := column(M,3);
```

We find a least-squares solution of $A\,x = b$ by solving the normal equations:

```
> M := evalm(transpose(A)&*A);
  k := evalm(transpose(A)&*b);
  x := matsolve(M,k);
```

The line is therefore $y = t + \frac{7}{5}$. Let's plot the line (in blue) on top of the data points:

```
> f := t -> t+7/5;
  p2 := plot(f(t),t=0..6,color=blue):
  display([p1,p2],view=[-1..6,0..7]);
>
```

The following plot shows the residuals. For each data point $[t, y]$, the "residual" is the difference between the actual y value and the value $f(t)$ predicted by the function f.

```
> residuals(f,data);
```

```
>
```

Finally, we calculate the error vector (also called the residual vector). Note that the components of this vector are the (signed) lengths of the vertical line segments in the above plot. So it is the sum of the squares of these lengths that we are minimizing when we find a "least-squares" solution.

```
> error := evalm(b-A&*x);
  evalf(mag(error));
>
```

====================

Now that we see the form that the coefficient matrix A and the right-side vector b must have, we can construct these matrices from our data set much more quickly. We first convert the data set into a matrix N whose first column contains the t coordinates of the data and second column the y coordinates. The vector b is the second column of N (i.e., the y coordinates of the data):

```
> N := matrix(data);
> tcoords := column(N,1);
> b := column(N,2);
>
```

Also, observe that A is the matrix whose first column is all 1's and second column is the first column of N (i.e., the t coordinates of the data).

```
> evalm(A);
```

To construct the matrix A the quick way, we apply the polyfit(..) command to the t coordinates of the data:

```
> A := polyfit(tcoords,1);
>
```

The above linear systems approach to the problem of fitting a line to a set of data points can also be used to fit other kinds of curves than just straight lines. In the next example, we fit a parabola to the same set of points as in Example 3A; in fact, a parabola produces a better fit.

====================

Example 3B: Fit a quadratic function $f(t) = a_0 + a_1\,t + a_2\,t^2$ to the data points in Example 3A:

```
> data := [[1,3],[2,3],[3,4],[4,5],[5,7]];
> f := t -> a[0]+a[1]*t+a[2]*t^2;
```

Solution: As in Example 3A, we have five equations, but now we have three unknowns: a_0, a_1, a_2. Here is the system of equations that we want the data to satisfy:

```
> eqn1 := f(1)=3;
  eqn2 := f(2)=3;
  eqn3 := f(3)=4;
  eqn4 := f(4)=5;
  eqn5 := f(5)=7;
>
```

We wrote these equations out as a reminder of the least-squares problem we are trying to solve. We will not use them in our solution. Instead, we'll use the faster method in which we use polyfit(..) to construct the coefficient matrix A. In the above system of equations, note that the coefficient of a_0 is 1, the coefficient of a_1 is t, and the coefficient of a_2 is t^2. Thus the matrix A must now have a third column, the column of the squares of the t coordinates of the data:

```
> N := matrix(data);
> tcoords := column(N,1);
> b := column(N,2);
> A := polyfit(tcoords,2);
>
```

The argument "2" in the polyfit(..) command says that the highest power of t that we want is 2. If we had used "3" in place of "2", A would have had a fourth column, the cubes of t.

We construct and solve the normal equations:

```
> M := evalm(transpose(A)&*A);
> k := evalm(transpose(A)&*b);
> x := matsolve(M,k);
```

Then we construct our best fitting quadratic function f(t) and plot it on top of the data points:

```
> f := t -> dotproduct(x, [1,t,t^2]):
  f(t);
> p2 := plot(f(t),t=0..6,color=blue):
> display([p1,p2],view=[-1..6,0..7]);
>
```

The quadratic fit is a clear improvement over the linear fit. We confirm by looking at the residuals and checking their sum of squares:

```
> error := evalm(b-A&*x);
> evalf(mag(error));
> residuals(f,data);
>
```

====================

In the Problems, we will fit other kinds of curves to data points.

Problems

Problem 1: Least-squares solution of $A\,x = b$

(a) Find the least-squares solution of the linear system $A\,x = b$, where A and b are defined below. Rather than using the normal equations, however, compute the projection of b onto the column space of A and use the projection vector to solve the linear system.
(b) Check your answer in (a) by solving the normal equations.

```
> A := matrix([[1,1,1],[2,1,0],[-2,-1,0],[1,0,1]]);
> b := colvector([2,5,4,1]);
>
```

Student Workspace

Problem 2: Compare linear and quadratic fits

Note: When finding the least-squares solution of $A\,x = b$ for this problem, use the efficient method of Example 3B for constructing the matrices A and b.

The data points $[t, y]$ below represent the heights in inches, t, and weights in pounds, y, of eight individuals.

(a) Find the straight line that best fits the data by solving the appropriate normal equations.

(b) Plot the line and the data points together.

(c) Find the quadratic function (i.e., parabola) that best fits the data by solving the appropriate normal equations.

(d) Plot the parabola and the data points together.

(e) Compare the sum of the squares of the residuals for the results in (a) and (c). Which curve gives the better fit?

```
> hw := [[64,141],[66,148],[68,157],[70,163],[72,172],
  [74,186],[76,206],[78,221]];
> p1 := plot(hw,style=point,symbol=circle,color=red):
  display(p1);
>
```

Student Workspace

Problem 3: Cubic curve fitting

Note: When finding the least-squares solution of $A\,x = b$ for this problem, use the efficient method of Example 3B for constructing the matrices A and b.

(a) Find the cubic function, $f(t) = a_0 + a_1\,t + a_2\,t^2 + a_3\,t^3$, that best fits the following data by solving the appropriate normal equations.

```
> data := [[2,9],[3,14],[5,25],[7,27],[9,36],[11,80]];
> p1 := plot(data,style=POINT,symbol=CIRCLE,color=red):
> display(p1);
>
```

(b) Plot the data points and the cubic curve together

(c) Explain where the linear system $A\,x = b$ comes from. Specifically, write the equation that comes from the first data point, $[2, 9]$, and explain what that equation contributes to the equation $A\,x = b$.

Student Workspace

Problem 4: Uniqueness of the least-squares solution

If the null space of A is $\{0\}$, prove that $A\,x = b$ has only one least-squares solution. (Hint: Any least-squares solution of $A\,x = b$ is also a solution of $A\,x = p$, where p is the projection of b onto the column space of A. Why does $A\,x = p$ have only one solution? **Warning:** Do not assume that A has an inverse; in least-squares problems, it almost never does.)

Student Workspace

Problem 5: Formula for the projection.

The projection of b onto the column space of A is given by the following formula (which is only valid when the null space of A is $\{0\}$):

$$p = A\,(A^T A)^{(-1)}\,A^T b$$

(a) Check the above formula for the linear system $A\,x = b$ in Example 1A (repeated below). That is, compute the projection p by this formula and also by projecting b onto an orthogonal basis of the column space of A.

```
> A := matrix([[1,2],[3,5],[-1,1]]);
  b := colvector([1,3,2]);
```

(b) Use the above formula to find the 3 by 3 matrix that projects vectors onto the plane $2\,x - y + 3\,z = 0$. Check your answer against the result of the projectmat3d(..) command below.

```
> projectmat3d([2,-1,3]);
>
```

(c) Derive the above formula for p. (Hint: The least-squares solution s of $A\,x = b$ satisfies both of the following two equations: $A\,s = p$ and $A^T A\,s = A^T b$. Put these two equations together. **Warning:** Do not assume A has an inverse; in least-squares problems, it almost never does. However $A^T A$ does have an inverse.)

Student Workspace

Problem 6: Fit a plane to multivariate data

We can also use our least-squares method to fit a surface to data points with three or more coordinates. For example, we can fit a linear function z = $f(x, y) = a_0 + a_1\,x + a_2\,y$ (i.e., a plane) to data points $[x, y, z]$. Consider the following data set, from which we want to predict the volume of useable wood produced by a commercially grown tree, knowing the diameter and height of the tree. The first coordinate, x, represents the diameter of a tree (measured in inches at a height of 4 feet); the second coordinate, y, represents the height of the tree in feet; and the third coordinate, the dependent variable z, represents the volume of useable wood in cubic feet.

```
> dhv := [[14.5,76.0,31.4],[14.2,75.0,29.9],[15.7,85.0,40.8],
   [14.4,76.0,31.0],[16.8,71.0,35.5],[20.5,82.0,65.8],
   [14.1,80.0,32.7],[14.8,69.0,31.3],[18.9,72.0,48.4],
   [14.0,66.0,25.5]];
>
```

Set up and solve the appropriate normal equations. Explain where the entries of A and b come from. Note: Fitting a linear function of more than one variable to a set of data points, as you did in this problem, is usually called "multiple regression."

Student Workspace

Problem 7: Log-log data

We can also fit power functions of the form $y = k\,t^m$ to data sets. However, this function differs in one important respect from the other function forms we have considered. In the other function forms, the unknowns were coefficients of known functions; in effect, we were seeking the unknown coefficients in a linear combination of known functions. In the function form $y = k\,t^m$, however, the unknown m is an exponent rather than a coefficient. To change this function form into one in which the unknowns are coefficients in a linear combination, we apply the natural log function to both sides of this equation.

(a) (By hand) Show that applying the natural log function to both sides of $y = k\,t^m$produces a linear equation of the form $Y = m\,T + b$. What are Y, m, T, and b (in terms of y, k, t, and m)?
(b) Johann Kepler (1571-1630) conjectured that the periods of revolution, y, of the six inner planets and the semimajor axes of their elliptical orbits, t, are related by a function of the form $y = k\,t^m$. Below is a data set $[t, y]$ for the six inner planets, where t is measured in days and y in millions of kilometers. Find the least-squares approximations to k and m as follows: Convert the equation $y = k\,t^m$ to the linear form $Y = m\,T + b$, and convert the data set as well; find the best fitting linear function to the converted data set; then find k and m for the power function.
(c) Plot the converted data set (which should look approximately linear) together with your best fitting linear function, and plot the original data set together with your best fitting power function.

Maple Note: Converting the Kepler data below is not easy to do; so it has been done for you. This transformation of the data is called the "log-log" transformation.

```
> Kepler := [[87.97,58.0],[224.7,108.0],[365.26,149.0],
  [686.98,228.0],[4332.59,778.0],[10759.2,1426.0]];
> M := matrix(Kepler);
> T := map(ln,column(M,1));
  Y := map(ln,column(M,2));
> N := augment(T,Y);
  LogK := convert(N,listlist);
>
```

Student Workspace

Pencil and Paper Practice Questions

Note: Maple is not required for any of the problems in this section. Do all of these by hand.

1. (a) Write the system of equations

$$x_1 = 1$$
$$2\,x_2 = 1$$
$$x_1 - x_2 = 3$$

in the form $A\,x = b$.

(b) Find the normal equations, and write them in the same form as the equations above (i.e., a system of three linear equations in two unknowns).
(c) Solve the normal equations by Gaussian elimination.

2. Suppose we want to find the line $y = m\,t + c$ that best fits the data points

$$[2, 5], [3, 4], [4, 2]$$

(a) Write down the resulting system of three equations in the two unknowns c and m. Then rewrite the system of equations in the form $A\,x = b$, where x is the vector with unknown components c and m. (Do not solve the system.)
(b) Give a geometric explanation for why the system $A\,x = b$ in (a) must be inconsistent. (Look at a plot of the three points.)

LAMP Installation Instructions

for Maple V Release 5 on a Macintosh

1. Double-click on the self-extracting archive lampR5.sea on the LAMP diskette. A window will come up that allows you to specify where to put the folder lampR5, which contains all the LAMP folders and files. We recommend that you put it in the top level of your main hard disk. Click on the Extract button when ready.

Aside: Within the lampR5 folder are the folders lampmod, which contains the modules, and lamplib, which contains the three library files, maple.hdb, maple.ind, and maple.lib.

2. Determine where Maple V Release 5 is installed on your system. If you are running Maple, quit it.

3. If you do not have a file called MapleInit in your folder for Maple V Release 5, copy the file MapleInit from the

installation diskette into that folder. (MapleInit is Maple's initialization file and tells Maple where to find the LAMP library.)

If you already have a file called MapleInit in that folder, copy the contents of the MapleInit file on the installation diskette to the end of your existing MapleInit file. (You can use SimpleText to do the copying.)

This completes the installation.

Special instructions:

A. If you did not put the lampR5 folder in the top level of a disk called "Macintosh HD," you must edit the line of MapleInit that says

libname := `Macintosh HD:lampR5:lamplib`, libname;

Change the path name to specify where you put lamplib. For example, if you put lampR5 in a folder called "Math" on a disk called "Public," the path name would be

`Public:Math:lampR5:lamplib`

NOTE: The quotes are BACK QUOTES.

B. If the Maple command with(lamp); does not bring up the LAMP library, type

libname;

in a Maple worksheet. The first name in the output should be Macintosh HD:lampR5:lamplib (or whatever folder you put the library in). If it is not, then Maple is not reading MapleInit. Check that you have put MapleInit in the correct folder.

If the first name in the output of libname is the correct folder name but Maple does not find the LAMP library, check that the folder name assigned to libname in MapleInit is where you put the three LAMP library files.

LAMP Installation Instructions

for Maple V Release 5 in Windows 95

Windows®

1. Double-click on the self-extracting archive lampR5.exe on the LAMP diskette. A window will come up that allows you to change where the archive intends to put the LAMP folders and files. However, we recommend that you either accept the default, c:\lampR5, or that you choose another easily accessible location

Aside: Within the lampR5 folder are the folders which contains the modules, and lamplib, which contains the three library files, maple.hdb, maple.ind, and maple.lib.

2. Determine where Maple V Release 5 is installed on your system. It is usually in the folder

c:\Program Files\Maple V Release 5

Within that folder, open the folder bin.wnt, which is usually the folder from which Maple is started. If you are running Maple, quit it.

3. If you do not have a file called maple.ini in bin.wnt, copy the file maple.ini from the installation diskette into the bin.wnt folder. (maple.ini is Maple's initialization file and tells Maple where to find the LAMP library.)

If you already have a file called maple.ini in bin.wnt, copy the contents of the maple.ini file on the installation diskette to the end of your existing maple.ini file. (You can use NotePad to do the copying.)

This completes the installation.

Special instructions:

A. If you did not put the lampR5 folder in c:, you must edit the line of maple.ini that says

```
libname := `c:\\lampR5\\lamplib`, libname;
```

Change the path name to specify where you put lamplib.

NOTE: The slashes are DOUBLE BACK SLASHES and the quotes are BACK QUOTES.

B. If the Maple command with(lamp); does not bring up the LAMPlibrary, type

```
libname;
```

in a Maple worksheet. The first name in the output should be c:\lampR5\lamplib (or whatever folder you put the library in). If it is not, then Maple is not reading maple.ini. Check that you have put maple.ini in the correct folder and that the Maple shortcut you use does run Maple from the folder bin.wnt.

If the first name in the output of libname is the correct folder name but Maple does not find the LAMP library, check that the folder name assigned to libname in maple.ini is where you put the three LAMP library files.